$\dfrac{1}{10}$

URAL OWL.
SYRNIUM URALENSE.

EX LIBRIS

*William A Hitch*

# BLAMING TECHNOLOGY

TECH

# Samuel C. Florman

# BLAMING
# NOLOGY

## The irrational search
## for scapegoats

ST. MARTIN'S PRESS
NEW YORK

*Design by Manuela Paul*

10  9  8  7  6  5  4  3  2  1
First edition


*Library of Congress Cataloging in Publication Data*

Florman, Samuel C.
    Blaming technology.

    1. Technology. I. Title.
T49.5.F56        306'.4        81-5818
ISBN 0-312-08362-9            AACR2

Chapters 2, 5, 9, 10, 11, 12, and 14 are based upon articles which have appeared in *Harper's*, as are portions of Chapters 8 and 15. Portions of Chapters 3 and 7 are adapted from articles published in *The American Scholar*; and Chapter 16 is based on an article which appeared in *Alternative Futures*.

FOR JUDY

Also by Samuel C. Florman

Engineering and the Liberal Arts:
A Technologist's Guide to History, Literature,
Philosophy, Art, and Music

The Existential Pleasures of Engineering

# CONTENTS

|   | Introduction | ix |
|---|---|---|
| 1 | Taken Aback in Michigan | 1 |
| 2 | Technology's Minor Moments | 11 |
| 3 | Technocracy: A Short, Unhappy Life | 23 |
| 4 | The Myth of the Technocratic Elite | 29 |
| 5 | Hired Scapegoats | 42 |
| 6 | Nuclear Angst | 52 |
| 7 | Muddled Heads and Simple Minds | 70 |
| 8 | Small Is Dubious | 80 |
| 9 | On-the-Job Enrichment | 97 |
| 10 | Codifying the Future | 105 |
| 11 | The Feminist Face of Antitechnology | 120 |
| 12 | The Gentle Sophistries of the Club of Rome | 131 |
| 13 | The Spurned Professional | 148 |
| 14 | The Image Campaign | 155 |
| 15 | Moral Blueprints | 162 |
| 16 | Technology and the Tragic View | 181 |
|   | Acknowledgments | 194 |
|   | Notes | 196 |
|   | Index | 199 |

# Introduction

Shortly after the publication of my book *The Existential Pleasures of Engineering* in late 1975, I received invitations to speak at a number of universities. Technology and its role in society seemed to be very much on everyone's mind, and lectures, seminars, and symposiums dealing with the topic were being scheduled at campuses all over the nation. Since I work full time as an engineer-contractor, I could not accept all of the invitations that were extended; but I did accept a few, first out of curiosity, and then because of the pleasure and stimulation that the visits afforded.

The college campus is a deceptively serene place, where jovial professors carry theories under their tweeds like stilettos, and students, after disarming a visitor with wide-eyed veneration, toss up challenging statements like cherry bombs. I rarely returned from an excursion into academe without a store of new thoughts and doubts. I have incorporated into the following pages many of the things I learned, as well as some of the things I said, at these campus gatherings.

At about the same time that I began my academic wanderings, the editors of *Harper's* asked me to write a series of articles, and subsequently to become a contributing editor. There is surely no better way to energize one's wits than by writing for a respected national magazine. With every aspect of the technological scene suddenly transmuted into potential material for an article, I found my intellectual antennae perpetually aquiver. From 1976 to 1980, I wrote a dozen pieces for *Harper's,* and some of these I have also adapted for use in this book. In my association with *Harper's,* Lewis Lapham, the editor, was–to use a trite phrase he would strike out in an instant–an inspiration. His unrelenting insistence upon honesty and independence, and his hearty abhorrence of pomposity, encouraged me to burrow for truth among the platitudes.

Wherever I encountered discussion about the role of engineering in society, I found platitudes aplenty, and they were all the more disturbing because of the passion and erudition in which they were so often enveloped. Much as I enjoyed my work for *Harper's,* and my sorties into academe, I became increasingly troubled by the irrationality that seemed everywhere to dominate discussion of technology.

In *The Existential Pleasures of Engineering* I observed that between 1950 and 1975 there had been a decline in the status of the engineer, and, concurrently, the emergence of an antitechnological backlash. These developments seemed to stem from the belated recognition that technological progress brought with it certain disagreeable side effects. It was my hope in 1975 that this antitechnological mood would soon abate, and a more realistic appraisal of the human prospect take hold, one in which the engineer would be seen as neither messiah nor archfiend.

To my dismay, feelings about technology have become even more frenzied. Distraction spreads, not only among latter-day Luddites, but among beleaguered scientists and engineers as well. Technology threatens to become in the 1980s what communism was in the 1950s, or even what witchcraft was in Salem in the 1690s—a word so steeped in emotional implication that its very mention drowns out the voice of reason.

My aim in this book is to bring common sense to bear upon some of the issues that, loosely connected, constitute the public debate about technology. I do not propose to do battle on behalf of the engineering profession. Nevertheless, implicit in all that I write is a personal belief that technological creativity is a wondrous manifestation of the human spirit.

# BLAMING TECHNOLOGY

# 1

# Taken Aback In Michigan

n the fall of 1979 the humanities department of the University of Michigan's College of Engineering sponsored a symposium entitled "Technology and Pessimism." The purpose of the event, as announced in the prospectus, was "to examine why and in what ways technological development has led to pessimistic assessments of the future." Just a few years ago such a statement would have made no sense. Today it sounds like the most appropriate theme that could possibly be selected for an academic conclave. ¶I was invited to participate in the opening session of the three-day program, and found myself paired with Melvin Kranzberg, professor of the history of technology at Georgia Tech and long-time editor of the quarterly *Technology and Culture*. Kranzberg spoke at 4:00 in the afternoon and I at 8:00. The following morning we were both scheduled to participate in a panel discussion. ¶If the planners of the event had hoped to start off with speakers who themselves felt pessimistic about technology, they had chosen

1

the wrong twosome. Kranzberg, joviality personified, quickly showed that he had no patience with prophets of technological doom. In ringing tones he reminded his audience that technology has progressively made life longer and more comfortable, while at the same time fostering the concept and practice of social justice. Problems have arisen, he maintained, because of peoples' continually rising expectations, and because in some instances technology has become appropriated by narrow interest groups. What is required, Kranzberg concluded, is social innovation–improved means of exerting democratic control over technological decisions.

When my turn came, I argued that neither childish optimism nor petulant pessimism was an adequate response to technological problems, and I suggested that a mature recognition of realistic possibilities would be more appropriate. Kranzberg had entitled his lecture "Technology: The Half-Full Cup." In the same vein, I told the parable of the two children in the garden, one who complained about thorns on the rose bushes, the other who said happily, "Mommy, the thornbushes have roses!" We were both so reasonable–"on the one hand, and on the other hand . . . ."–and the reception accorded us was so friendly, that it was difficult to imagine there being a single person in the audience who did not agree with what we were saying.

The mellow mood that prevailed reminded me of a program in which I had participated a couple of years earlier, a symposium at Lafayette College entitled "Technology and the Human Condition." On that occasion Isaac Asimov had entertained one and all with his mockery of antitechnologists who would like to be back in old Athens "yakking it up with Socrates." The truth is, said Asimov, that most of the people in ancient Greece worked like animals and died young. We should be thankful for technology; it gets a bad press; it can be used wisely or unwisely. All of the other speakers agreed, the audience applauded, and everybody went home in high spirits.

This is just like it was at Lafayette, I thought to myself, as I got ready for bed that evening in Michigan. Everything so cordial and civilized. Thank goodness it isn't like Albion College, where I had to debate lifestyles with a hippie from a commune and an abbott from a monastery, nor like Smith College, where an angry feminist reviled technology and its defenders for being "macho." I enjoyed the deep sleep of the complacent.

The next morning's panel discussion was moderated by John Broomfield, a University of Michigan history professor. Since I had enjoyed his affable company at a faculty dinner the previous evening, I was startled when he opened the proceedings on a hostile note. "No shred of optimism," he said, "is added to my view of our current technology and its spokespeople by the arguments of our first two speakers." He then launched into a bitter tirade against high technology and its sponsors. I had heard its like before, but I found this version to be particularly unsettling because it was so unexpected. Technocrats, according to Broomfield, are developing large-scale technological systems that disenfranchise the average citizen, erode civil liberties, and produce "specialized ignorance for some and generalized ignorance for most." This new order is perilously vulnerable to disaster because, being exceedingly complex, and neither "natural" nor "biological," it has become "intolerant of mistakes." Not only is high technology the malignant product of a bureaucratic technocracy, Broomfield continued, but it also appears to have gotten out of control: ". . . if you have a new technique you encourage its widespread adoption; if you have a new product you peddle it everywhere . . ."

The professor's attack encouraged others in the audience who had been silent the previous day. Soon we were hearing a young man compare farm tractors unfavorably with oxen; a woman rose solemnly to announce that the very building in which we were meeting should never have been constructed since it was situated on terrain that once was sacred to a local Indian tribe. By the end of the morning session, my mood had soured considerably.

A few days later, an article about the conference appeared on the front page of *The New York Times* science section. At first I was pleased to see it; there is a primal satisfaction connected with finding one's name in the newspaper. But then I wondered, considering all the important developments in science and technology, why the dispirited musings of an academic symposium should be receiving so much attention.

"Scholars Confront the Decline of Technology's Image," read the headline. This reminded me of how often in recent months I had seen similar headlines prominently displayed in this same newspaper: "Skylab and Other Mishaps Tarnish Technology's Image," "Scientific Decisions: The Baffled Public," "Scientists and Society's Fears," and so forth. In a series of philosophical essays, this august voice of the American establishment had been brooding about the public's fear of technology. As I read the article about the Michigan symposium, I felt increasingly uneasy. Academic handwringing is one thing, but for the press to treat it as important news is something else entirely.

To be sure, in recent months there had been no shortage of anxiety-producing technological incidents: oil spills, airplane crashes, collapses of dams and auditorium roofs, the descent of Skylab, the discovery of toxic materials at Love Canal, and the uniquely alarming accident at the Three Mile Island nuclear plant. Yet reciting this litany does not explain a change in the journalistic perception of technology. For never in the history of newspapers has there been a shortage of technology-related catastrophes. They have long been the very stuff of headlines and extra editions: train wrecks, falling bridges, bursting boilers, collapsing buildings, devastating fires, explosions in mines and factories, lead poisoning, botulism. . . . Incidents today occur no more frequently than they used to, and the consequences are no more ghastly. There were also, in the past, plenty of long-range, enduring torments associated with technology–city slums wretched beyond telling, disease-causing tanneries, foul-smelling gas works, grime-filled air, and more–a long, sordid list we have been happy to forget.

Nor was there ever a shortage of spirited reporters and editors to report on these intolerable conditions. It was always the pride of journalists to complain about carelessness, ignorance, laziness, and greed, and to insist that in the future there be improved performance. Even the revelations that led to identification of an "environmental crisis" were greeted at first with outrage and sermonizing. Only recently has indignation seemed to be giving way to malaise. When newspapers start becoming melancholy–or even dwelling upon the melancholy of others–one cannot help thinking that something is terribly wrong. At least that is the way I felt as I read in the *Times* of "growing doubts about society's ability to rein in the seemingly runaway forces of technology. . . ."

What is it, however, that is really running away? Is it technology, or is it fear of technology? The newspapers print scare stories; people read the stories and become alarmed; and then the newspapers quote the people expressing their alarm. Television, with reports that stress helplessness in the face of calamity, increases the tension. An academic symposium is scheduled to consider what is happening, and it in turn becomes news. The growing hysteria feeds on itself.

Perhaps this is not being fair to the media. The current mood of apprehension cannot have been created out of thin air by editors and reporters. It has percolated through our social consciousness in a diffuse pattern that is almost impossible to trace. It starts with disasters, but as I have said, there have always been disasters. It is the way we respond to disasters that has changed. If occasional indignation has given way to lingering skepticism, that is understandable enough. Utopia was promised, and it isn't coming. But if a sense of helpless resignation is taking over–if technology, after changing from boon to disappointment, is now perceived to be changing from disappointment to *threat*—then somebody had better try to find out what is happening, and quickly.

Clearly, the change in mood has been inspired in part by a number of lionized science writers such as Rachel Carson and Barry Commoner. These persuasive alarmists have cautioned that the problems we face today are far more grave than those we faced in the past. Whether or not this is true, these critics have given useful warning and gone on to suggest alternative courses of action. It cannot be said that they have tried to cultivate fear of technology itself. Yet their eloquent polemics may have oversensitized the public in ways that were not intended. The same paradox pertains to advocates of the "counterculture" who, in voicing opposition to certain manifestations of technology, have helped to spread the belief that "the machine" will inevitably take over. Warnings intended to bring about constructive action seem inadvertently to be spreading fear and paralysis.

Another source of the new anxiety lies deep in the tomes of such introspective scholars as Jacques Ellul and Lewis Mumford. Not many people buy, much less read, the works of these savants, yet their dolefully deterministic view of technology is revealed to the public in book reviews, and disseminated throughout intellectual circles by articles in abstruse journals. Starting in the early 1970s, the concept of runaway technology became much the vogue among academics. Articles, books, and dissertations devoted to this theme began to appear in great number, abetted no little by funding of "technology and society" studies by both the National Science Foundation and the National Endowment for the Humanities. By 1977 this school of literature was growing at such a rate that Langdon Winner, an MIT professor, was moved to review the issue in a book entitled *Autonomous Technology: Technics-out-of-control as a Theme in Political Thought.*

Book titles tell a lot about what is happening to a culture, and during the 1970s there was ample evidence that the American love affair with technology was in trouble. Bookstore windows displayed serious works with titles like *The Illusion of Technique* and

*The Poverty of Power,* along with stacks of mass-market paperbacks like *Future Shock* and *Overskill.*

Such subliminal influences as book reviews and bookstore windows have been reinforced by the comments of respected sages as reported in important places. Robert Penn Warren, interviewed in *U. S. News & World Report,* predicts that "in the technetronic age . . . the boys who handle the postcomputer mechanisms . . . will inevitably be in control . . . with a vast, functionless, pampered and ultimately powerless population of nonexperts living on free time, unemployed and unemployable." [1]

Daniel J. Boorstin, Librarian of Congress and Pulitzer-Prize-winning historian, writes an essay for *Time* magazine striking a somber note: "The Republic of Technology where we will be living is a feedback world. There wants will be created not by 'human nature' or by century-old yearnings, but by technology itself." [2]

John Hersey speaks at a convocation at MIT, and his remarks about "the growing public hostility to technology" are reported in the syndicated column of Anthony Lewis.[3]

Antitechnology, which for a while seemed to be a rather harmless—possibly even wholesome—undercurrent of intellectual rebellion, is suddenly a rushing tide. In their comfortable parlors, readers of *The New Yorker* are subjected to weekly elegies: "The Faustian proposal that the experts make to us is to let them lay their fallible hands on eternity. . . ." [4] In thousands of barbershops, the readers of *Penthouse* are given less subtle warnings: "In the darker sense, technology is *whatever you're not supposed to understand.*" [5]

A measure of the pervasiveness of the new mood is the consternation evinced by once-haughty scientists and engineers. In the spring of 1980, Philip Handler, president of the National Academy of Sciences, writing in the prestigious journal *Science,* expressed his concern in uncharacteristically urgent tones:

The intellectual elite in every era has always been pessimistic. But today, concerned that "that which can be done, will be done," there has arisen an antiscientific, antirationalistic trend that should give us pause. . . . That antiscience attitude perniciously infiltrates the news media, affecting the intelligentsia and decision-makers alike. It must be confronted at every opportunity.[6]

Apparently the seriousness with which *The New York Times* regarded the symposium on "Technology and Pessimism" was both effect and cause, a symptom of the nation's growing concern as well as a source of new anxiety.

Many intelligent and responsible people, however, seem not to be unduly alarmed about technological change. And public opinion polls show that the average citizen still considers science and technology to be, on balance, forces for good. Maybe I am overreacting.

In my previous book I carefully considered, and tried to rebut, the antitechnological arguments of Jacques Ellul, Lewis Mumford, René Dubos, Charles Reich, and Theodore Roszak. A reviewer, writing in *The Chronicle of Higher Education,* said that for an engineer to complain about the works of such people "seems a little like an elephant complaining of being bullied by gnats." [7] Well, perhaps—but then it is common knowledge that gnats can torment large animals to distraction. A sting here, a buzz there, and all of a sudden we have the beginnings of a calamitous stampede. There is small comfort in thinking of the antitechnologists as gnats.

There are many different ways of trying to understand the antitechnological movement. A theologian might conclude that it is a long overdue recognition of the ebb of rationalism, a development that could have been predicted four centuries ago when Copernicus and Galileo first challenged the teachings of the Church. A political scientist might view the changing mood as an expression of incipient revolution, since technology is perceived by some to be the means by which the ruling classes perpetuate their power over the masses. Through an anthropologist's prism, antitechnology can be seen as a new mythology, welling up in the effervescent depths of the cultural subconscious.

If one thinks in the language of psychology, a completely different hypothesis comes to mind, namely that the dread of technology is nothing other than a phobia, that condition which develops when people *displace* their anxieties (about death, separation, parental rage, or whatever) in an effort to control them. It is well known that certain objects or situations–such as crowds, heights, closed spaces, open spaces, and a variety of animals–can be endowed by individuals with special meaning, and thereafter regularly induce a reaction of anxiety. Omnipresent technology is an apt candidate for this list. Fear of flying is, in fact, being treated as a phobia, and much publicity has been given to the cures effected by techniques of behavior modification. The anxieties of people living in the vicinity of the Three Mile Island nuclear plant have also been the subject of much interest and some study.

Another way of defining the antitechnological movement is as a school of literary expression. According to Leo Marx, a leading light of MIT's Program in Science, Technology, and Society, the current pessimism about technology is a renewed manifestation of *pastoralism,* "an ancient mode of thought and expression embodying a negative response to social complexity and change." [8] Pastoralism, according to Marx, characteristically flourishes in times of accelerating social change, Rome in the time of Virgil, for example, or England during the Elizabethan Age, or, in our present case, the United States since the start of the industrial revolution.

To the antitechnologists, of course, all of these theories are ridiculous or, at best, beside the point. In their view, historical perspectives no longer apply, since changes have occurred that are unprecedented and fundamental. They see no sense in harking back to Virgil or the industrial revolution when we live in an age of computers, recombinant DNA, and nuclear power (to say nothing of nuclear weapons), with our resources dwindling and our environment degraded. If anyone is theologically confused, mythically muddled, or neurotically disturbed, they believe it is those of us who cannot see that technology really is out of control, and that technocrats really are creating a hell on earth.

I am convinced that the antitechnologists are misguided. But the intensity of the antitechnological argument persuades me to pause. Too much of the discussion about technology, pro and con, has been on an intuitive level (where one can convince only oneself), or else cloaked in complicated theories that arise less from evidence than from the majestic intellects of their creators. Perhaps we can find some simple facts that will persuade an open-minded reader that the antitechnologists have been listening to the beating of their own hearts instead of looking at the world around them. It is easy to *feel* that technology is out of control and/or that technocrats are leading us astray. Such feelings appear to be spreading. They make for attention-getting academic symposiums and mournful essays by science reporters. Yet I believe it can be shown that technology is still very much under society's control, that it is in fact an expression of our very human desires, fancies, and fears.

# 2
## Technology's Minor Moments

Whatever happened to. . . ? Something that once filled the stage of our daily lives vanishes so completely that we are amazed, and we turn our confusion into a game. We call it Trivia. ¶Lately, I have found myself asking this question about technologies. An automobile commercial appears on television and I wonder: Whatever happened to the rotary engine? I see a scuffed shoe and ask: Whatever became of Corfam?

So many wondrous inventions, materials, products—entire technologies—that were the center of attention just a few years ago seem to have dropped from sight. For a while new technologies were appearing in such profusion that many intellectuals feared the process had escaped from human control. Even those observers who rejected deterministic theories had to admit that the phenomenon was somewhat unsettling. Inventions seemed to take on a life of their own, dependent more upon the cleverness of the product than upon any demonstrated human need.

But now they are gone—some of the very technologies that most astonished us—and I find that the occasional recognition of their absence is a peculiarly intense experience. I feel like the Proustian narrator who steps on an irregular flagstone and is suddenly flooded with memories of Venice, except that my mindscape is filled with engines, waves, wires, and miraculous materials.

Recalling failed technologies would be no trivial pastime if it served to show how wrong is the widespread notion that technology conquers all.

I am not referring to failures of style, like the Edsel, nor to the myriad everyday mishaps that are endemic to technological development (and are a necessary part of technological advancement), nor even to those intriguing failures that are of special interest in scientific and engineering circles. The Institute of Electrical and Electronics Engineers devoted the October 1976 issue of its journal *IEEE Spectrum* to the topic, under the headline "What Went Wrong?" Most of the failures they discussed, however—automated zipcode-reading equipment, 3-D radar for air-traffic control, thermoelectricity, two-way cable TV—were scarcely known to the general public.

Nor do I refer to technologies that are familiar only through anticipation, items such as moving sidewalks, plastic houses, and trains that travel at 200 miles per hour. These developments were among those promised to us by the editors of *Changing Times* when, in 1961, they looked ahead to the world of 1975. Other experts envisioned imminent breakthroughs in mechanized super-markets, farming and mining in the oceans, and roofed-over cities. Such forecasts tell us more about the black arts of futurism than about the success or failure of particular technologies, which cannot be said to have failed if they have not yet been seriously tried.

No, I refer only to those technologies that actually existed and mesmerized us all, launched on a flood of press releases, advertising, and go-go financial speculation, seeming to create their own markets and even to evoke appetites that had not existed before.

It may be thought anachronistic to label as failures those brilliantly creative engineering achievements which happen to become commercial disasters, particularly when many of them are likely to reappear sometime in the future. But, when industry has gone all out to nourish a product–spending millions, risking reputation, and rousing the public with the vaunted forces of American salesmanship–only to give up and acknowledge defeat, then it is reasonable to say that the product has failed. Further, since industrialists are held by their critics to be the greedy handmaidens of a demon technology, when they abandon a new product on which they have invested a bundle, the myth of invincible technological advance is itself discredited.

One of the most spectacular technological debacles of recent years has been the rotary engine, which hummed its way into our subconscious on the wings of a sales campaign launched by Mazda Motors in 1970. When the first rotary-engine Mazdas went on sale in California, 86,000 people poured through 27 showrooms in three days. The engine was a consumer's delight, as well as an engineering triumph.

The inside of a rotary engine looks incredibly simple–a triangle rotating within an oval casing. (More precisely, the casing is a two-lobed epitrochoid; the beauty of the invention lies in its ingenious geometry.) The spaces between the outside of the triangle and the inner surface of the casing act as moving chambers of variable size. As ignition occurs on one side of the triangle, the second side, turning, forces exhaust out through an opening in the casing, while at the third side air and fuel are being taken in and compressed. The spinning triangle is attached to the end of the crankshaft. This arrangement is in marked contrast to the conventional automobile engine in which a number of pistons go up and down through a four-stroke cycle, transmitting power to the crankshaft through individual connecting rods.

The rotary also differs from the conventional automobile engine in that it has no valves. The fuel–air inlet port and the exhaust port are opened and closed at the appropriate time by the passage of the triangular rotor itself. Less vibration, less noise, less than half as many parts as a comparable V-8 engine, and less than half the weight–a combination of improvements almost too good to be true. Mazda's parent company in Japan, Toyo Kogyo, had been perfecting the engine for a decade, ever since negotiating a license in 1960 with its German patent-holders. Nearly 500 design modifications had been evaluated, and 5,000 engines tested.

Admittedly, engineers had not quite solved the problem of maintaining a satisfactory seal where the rotor comes in contact with the casing, and the consequent slight leakage, along with comparatively low combustion temperatures, resulted in disappointing fuel economy and excessive emission of pollutants. But in 1970 this flaw was thought negligible.

Soon after the first Mazdas appeared on American highways, General Motors decided to develop a rotary engine of its own and agreed to pay $50 million for nonexclusive rights to the existing patents. The license agreement was signed November 11, 1970, a date that promised to loom large in the history of technology. A year later, according to the *Chicago Tribune,* the top people in the automotive industry were predicting that by 1980 there would be "a complete changeover" from reciprocating piston engines to the rotary type. Science writers of the nation were enthralled. *Popular*

*Science* featured a "Sneak Preview!" of GM's new rotary engine for the 1974 Vega; *Scientific American* ran a feature article on the engine; *Fortune* announced that the engine heralded "a new automotive age"; and *The New York Times* profiled its aging inventor, Felix Wankel.

The stock market reflected the public's enthusiasm. Curtiss-Wright Corporation, a deficit-plagued aeronautical manufacturer had, in 1959, acquired exclusive North American rights to the engine, and so was a beneficiary of the General Motors licensing deal. During the first half of 1972 Curtiss-Wright common stock rose more than 46 points—from 13 to over 59—and a *New Yorker* cartoon showed an old man asking his broker, "Do you think I might have time to make it big on the Wankel engine?" In the toy market, models of the engine were a big seller.

In 1973 Mazda sales reached 119,000 cars, and the company's advertising budget approached $15 million. The piston engine, mocked in animated commercials for going *boing-boing* instead of *hmmmm,* appeared on its way out. The first GM rotary models were awaited eagerly. Ford and American Motors hurriedly signed license agreements.

By mid-1974, however, things had changed. The OPEC oil embargo suddenly made fuel economy—the rotary engine's weak point—critically important. Mazda sales dropped precipitously, and Ford shelved its rotary project after having spent more than $10 million ("A total waste," said Henry Ford II). The *Wall Street Journal* reported that the rotary had become "a sputtering engine." But General Motors was not ready to give up. In addition to having already paid $40 million of its $50 million licensing fee, the company was estimated to have spent between $250 million and $300 million on development costs. According to the automotive editor of the *Christian Science Monitor,* this was "too much money to simply write off—even for General Motors."

By the end of 1974, however, GM had decided to "postpone" introduction of its rotary engine. In addition to the fuel-economy problem, it was found that hydrocarbon emissions continued to exceed anticipated federal standards. Finally, in April 1977, swallowing its pride and a loss of millions, GM quietly closed down its rotary engine development program.

General Motors is not the only corporate giant to have been embarrassed by a highly touted technology that failed. Corning Glass Works was among those companies that invested heavily in an ill-fated development called fluidics. This mellifluous term, which in 1967 *Popular Mechanics* called "the next household word," made its public debut that same year in a full-color feature in *Life*. The visual presentation was lavish, and the text was exuberant: "Already hundreds of U.S. Companies have gambled millions of dollars on the future of fluidics . . . an overdue idea."

The concept of fluidics was appealing–using liquids or gases to perform functions, such as switching or amplification, that are ordinarily performed by electronic devices. It is a characteristic of fluid flow that any liquid or gas–say, a jet of air–moving through a tube will, if it comes to a Y-shaped fitting, cling to one surface of the tube, and thus travel down one arm or the other of the Y. If a second small jet (the "control jet") is injected at the neck of the Y, it can, with a small nudge, make the first jet (the "power jet") jump to the opposite arm of the Y and back again. Thus, one has created a simple switch device with no moving parts. If an abruptly enlarged chamber is provided at the neck of the Y, so that the air cannot cling to a side surface, then the jet will tend to flow equally into both arms of the Y. In this case, a small jet injected from the side can regulate the amount of flow into each of the arms. The device can now perform functions of proportional control and amplification. By building networks of tubes, or channels in stacked laminates, with an assortment of intersections and chambers, fluidics engineers created ingenious control systems, which, for a while, appeared likely to prove superior to their electronic counterparts.

Manufacturers set to work etching delicate patterns into glass, metal, and ceramic blocks, and producing plastic networks by injection molding. But they were doomed to frustration. No matter how they struggled to miniaturize their product, they could not keep pace with electronics engineers, who were moving into a microscopic realm. Transistors became incredibly tiny and even more incredibly inexpensive. Also, alas for supporters of the new technology, electricity moves at almost a million times the speed of air jets. Unable to compete in size, price, or speed of operation, fluidics became virtually obsolete almost before its boom could get underway—not quickly enough, however, to save a lot of people from losing a lot of money.

While the rotary engine succumbed to a suddenly changed environment, and fluidics was done in by the rapid evolution of a competitive technology (both reminiscent of the problems once encountered by the hapless dinosaur), sponsors of other failed technologies have no comparable excuses. Xerox, for example, assuming that people would pay a high price for rapid transmission of mail and other documents, invested heavily in a machine called the LDX (for long-distance xerography). In the mid-sixties, electronic facsimile transmission (called *fax* by knowledgeable investors) appeared to be a technology whose time had come. But none of the devices produced through the end of the Seventies—some of them marketed at great expense and with much hoopla—appealed to the average small business, much less to the average homeowner. The failure of fax to establish a mass market shows that there are limits to our sense of urgency. Eventually, a smaller, speedier, and cheaper fax machine will be developed successfully, and perhaps— the experts are ever more cautious with their predictions—the gadget will become as widespread as the telephone. Until that time, if we are willing to wait for the mailman, we are apparently not in as much of a hurry as some Xerox executives assumed.

Nor is our love of sensational gadgetry without its reasonable limit. The 3M Corporation spent a lot of money to develop a thin loudspeaker, only to discover that hardly anybody cared about having skinny hi-fi equipment. The Polaroid company was discomfitted to learn that most buyers of movie film are not willing to spend extra money to have it instantly developed. It was thought that holography, which re-creates objects three-dimensionally in space before our incredulous eyes, would enjoy widespread use in advertising, sales displays, instructional systems, 3-D movies, and many other fields. Du Pont and the Battelle Memorial Institute poured a lot of money into this technology, only to find out that, for now at least, holography seems to be a spectacular oddity that cannot make its way in the marketplace.

The du Pont name is associated with another famous loser, the miraculous material Corfam, which, instead of taking American consumers by storm, served to convince people how much they preferred natural leather.

Other wizards of alchemy have been similarly disappointed. Titanium, a metal with a high strength-to-weight ratio (and a name that sounds at once like a giant and a fairy queen), was for a while touted as the ideal material for almost every conceivable product. Except for such products as airplanes and rockets, however, its special attributes did not offset the high cost of obtaining it and shaping it, so with the mid-1970 slump in the aerospace industry, the titanium boom abruptly ended.

These are a few of the technologies that flash into my mind at unexpected times. There are others—ground-effect machines (vehicles that run suspended on magnetic fields or cushions of air), factory-built housing, automated mass transit systems, ultrasonic dishwashers, thermography, super-conductivity, weather modification—so many disappointments, and so much money lost by so many smart people.

Most of these technologies have come across the desk of Walter Cairns, manager of the new product venture arm of Arthur

D. Little, the well-known, Cambridge, Massachusetts, think tank. It is Cairns's job to select inventions that can be developed profitably, and he ruefully admits that he has been carried away, on occasion, by glamorous technologies that never lived up to their press notices. "A technology comes along," he says, "which is so darned fascinating that you just can't believe that it isn't important. Also, technologies become fashionable. Someone says the magic words, 'future growth area,' and everyone wants to become involved."

Overall, Cairns's record has been one of success, achieved through hardheaded insistence on a likely payoff. Out of every thousand ideas proposed to Arthur D. Little, from sources both within the organization and outside it, Cairns and his staff select only 60 as worthy of development. Out of these, only 20 find manufacturing companies willing to invest seriously in their commercialization, and within four years about ten of these 20 are likely to be abandoned. This means that of the proposals originally evaluated, only about 1 percent are destined for success. Viewed from Cairns's office, technologies seem anything but the implacable force they are often assumed to be. On the contrary, they seem frail and vulnerable, like performers waiting anxiously in a theatrical agent's anteroom.

Many amazing things can be done, but so what? Are they things that the public wants? Can they be done for a price that the public will pay? Engineering has been defined as doing for $1 what any fool can do for $2. The definition could be amended to specify that even the $1 must be a price that will attract a buyer.

The system, of course, does not work neatly in accordance with classic economic theory. Between the inventor and the public stands the corporation, preventing a pure interchange between creative skills and consumer wants. It is often said that big businesses deliver only those products that promise easy profits, slyly withholding many products that would benefit the public. But when looked at closely, corporate distortion of the ideal supply-

and-demand graph comes less from canny greed than from human frailty. The "calculating" executive is more likely to be thinking of his own problems than of the ways in which his company can bilk the public.

Myron Tribus, director of the Center for Advanced Engineering Study at MIT, whose career has included service as Assistant Secretary of Commerce and as a senior vice-president of Xerox, speaks knowingly on the subject: "In moving innovative ideas to production, the answers to human questions are often more critical than the technical problems. Who may lose position within the firm? Who will have to change work habits? Who will appear to be a hero? Whose idea will be proven false? Who risks reputation in backing the idea?"

More often than the public imagines, the fate of a new technology depends upon the erratic behavior of the people who make their living as corporate executives. Some observers of the automotive industry make much of the fact that one of the most ardent admirers of the rotary engine was a professor of engineering at the University of Michigan, whose father happened to be for a time president of General Motors. The president's resignation was followed by GM's abandonment of the engine. Every technological failure, like every success, is surrounded by a cloud of intrigue. A gossip column devoted to the shadowy world of the executive suite would do much to disprove the theory that technologies have a life of their own.

Stories of failed technologies are seldom told in the mass media. Success is what sells—along with disasters (which are usually accidents rather than evidence of failure). Nurtured on a diet of moonshots and oil spills, pocket calculators and power blackouts, we begin to feel that we live in a world of genies. (Who will do a study of the vast influence surreptitiously wielded by science writers?)

Yet wherever one really looks at technological development—among inventors, producers, distributors, or consumers—there is the human spirit at center stage, foolish, perhaps, as often as clever, dull as often as radiant, but undeniably, irrepressibly, vitally human.

If we cannot put our creations back in the bottle, we can at least continue to bring to bear upon them the discrimination that is uniquely ours. Even successful technologies are subject to human control. They can be discarded when people get bored (the once-ubiquitous CB radio, for example), restricted when people become concerned (the supersonic transport plane), or suppressed when people become resolute (the refusal of Congress to allow the Internal Revenue Service to establish a nationwide computer network for the retrieval of personal information).

Certain inherent qualities of the natural world suggest that some technologies will flourish without regard to personal choice. Electronic communication, for example, is becoming so cheap, and travel is becoming so expensive, that futurists envision a world in which much human contact will take place by means of glowing screens. So far, however, this trend is being resisted. Business executives who could rent television conference rooms at their local telephone company offices seem to prefer getting up before dawn and flying halfway across the country to meet face-to-face in motel rooms. Students still choose live classes over pretaped televised lectures. *Homo faber* will take what the universe gives, so to speak—from rivers and fertile farmland to electricity and silicon chips. But the life force has desires of its own, and will not easily be intimidated.

The large chrome-bedecked automobile, turned out in Detroit by the millions, is often used as an example of the mindless march of technology. Now that American consumers have shown their preference for small foreign cars, it is being said that the American automobile manufacturers "guessed wrong." *Guessed wrong!* How far away can one get from the concept of autonomous technology?

When I encounter a tale of technological failure, or suddenly recall that a brilliant engineering achievement has come to naught in the marketplace, my reaction is ambivalent. After the initial surprise and disappointment, there comes a sense of satisfaction. If promising technologies can suffer fatal blows from unexpected circumstances, if they are buffeted in the world along with other ideas and social forces, then there can be no technological imperative. This means that we are still—however precariously—in control of our own destiny.

# 3
## Technocracy: A Short, Unhappy Life

Beyond the idea of autonomous technology lies the concept of *technocracy*. According to this view, scientists and engineers are able to translate their special knowledge into power over their fellow citizens. ¶Ironically, the idea of ordering society in accordance with scientific and technical principles, although discussed from time to time by a few scientists and engineers, has appealed especially to nontechnical, humanistic intellectuals. Most of these individuals have represented the antithesis of everything that comes to mind when one thinks of engineers. ¶Traditionally, the roots of technocracy are traced to Henri de Saint-Simon, a French count who, after making a fortune in land speculation during the French Revolution, lavished his wealth on a salon for scientists. Saint-Simon conceived of a quasi-religious, socialist society in which science and technology would be applied to the solution of social problems. In this new social order scientists would take the place of priests. His ideas, expressed in a series of fervid books and pamphlets, appealed to a number of idealistic contemporaries, and when Saint-Simon died in 1825, a band of disciples carried on his teachings. The most important of these were two bankers, Benjamin-Olide Rodrigues and Barthélemy-Prosper Enfantin. They formed an association, and soon crowds flocked to hear lectures on "the Saint-Simonian faith."

In 1830 the Saint-Simonians issued a proclamation demanding ownership of goods in common and abolition of the right of inheritance. Members of the organization moved to Enfantin's estate, where they lived in communistic style, wearing blue tunics and red berets. Within a few years, after some of the leaders were convicted of offenses against public order and morality, the movement broke up.

I relate these events not in an attempt to discredit Saint-Simon (in whom Friedrich Engels found "the breadth of view of a genius"), but to demonstrate that the earliest call for a technocracy was not in the least scientific or technological. The Saint-Simonians paid homage to the concept of scientific rationality, but essentially they were romantics who followed the promptings of their hearts.

In the United States, the technocratic movement began with the ideas of Frederick W. Taylor (1856–1915), founder of scientific management, whose time-and-motion studies revolutionized American factory production. At about the time of World War I, a few self-proclaimed "progressive" engineers got together and discussed the possibility of applying Taylor's principles of scientific management to the reorganization of society. In 1919 Thorstein

Veblen, a radical economist who taught at the New School for Social Research in New York, met with some of these engineers and spoke of recruiting a "soviet of technicians." Veblen's ideas were set forth in his book *The Engineers and the Price System*.

But this was little more than idle speculation among a handful of people. Certainly most engineers had never heard of Veblen, nor of the few "progressives" who had met briefly in New York. The concept of technocracy languished, and during the booming 1920s there was little reason to talk about putting society into the hands of engineers. Society seemed to be doing well just as it was.

With the coming of the Great Depression, however, public complacency was shattered. Suddenly nothing seemed to be working and nobody seemed to have any answers. The stage was set for popular recognition of the technocratic idea. What occurred, however, was not a reasoned presentation of technical proposals by responsible people, but rather an incredible series of farcical episodes. Today it would be called a media event.

The tragicomedy began in the summer of 1932, when Walter Rautenstrauch, chairman of the department of industrial engineering at Columbia University, along with several engineers, social scientists, and self-styled experts of various sorts, founded the Committee on Technocracy. This group proposed to undertake a massive "energy survey of North America." A question that troubled all thinking people at the time was how a land with abundant natural resources, efficient industrial plants, and a willing work force could have stumbled into financial disaster. The survey undertaken by the committee was intended to evaluate American industry in terms of energy and production and to show—what should have been obvious—that since there were ample resources in the nation, it was the system that had broken down. Although the technocrats did not propose any specific remedy, they did imply that a "scientific" restructuring of the system was required, and that engineers, rather than businessmen and government officials, were the people most qualified to supervise this undertaking.

Rautenstrauch convinced Nicholas Murray Butler, then president of Columbia, to permit the committee's work to proceed under the auspices of the university, and during the fall of 1932, word circulated that a group of engineers at Columbia were working on a "solution" to the Depression. The press picked up the story, and with incredible speed a technocratic "movement" spread across the nation. Under sudden pressure to come forth with detailed strategies, some members of the Columbia committee made hurried and ill-considered proposals. One of these plans recommended distribution of "energy certificates" in equal quantities to each citizen, through which the allotment of goods would be managed.

It did not take long for intelligent observers to note that the technocrats' proposals were fatuous, and by January 1933, the "technocraze" was ridiculed. The Committee on Technocracy, disavowed by an abashed Columbia administration, broke up in confusion and acrimony.

A mass movement had been spawned, however, and political groups calling themselves technocrats were active, particularly in the western part of the nation, through 1932–33, claiming as many as 250,000 members. But within another two years, just a few diehards were left, and technocracy came to be thought of as a crank doctrine—one of many that thrived briefly during the worst days of the Depression.

As with the Saint-Simonian movement, American technocracy had nothing to do with engineering. Walter Rautenstrauch was an engineer, to be sure, but his specialty, industrial engineering, has always been on the periphery of the profession, blending into what today is called business management. Also, having spent 40 years at Columbia, Rautenstrauch could as accurately be classified an academic as an engineer.

The other leader of the original Committee on Technocracy—and the person most intimately identified with the term *technocracy*—was a strange man by the name of Howard Scott. Scott was not an engineer. He was a pseudo-philosopher with a mysterious past who appeared in Greenwich Village in 1918 and excelled at holding forth in coffeehouse discussions. Although Scott made his living selling floor wax, and had received no formal professional education, he claimed to be an engineer. The eventual exposure of this fraudulent claim was partly responsible for the downfall of the Committee on Technocracy. In the waning days of the technocratic movement, Scott took to wearing an all-gray uniform and accepting the salutes of his aides.

Engineers were much sought after by the technocrats, but few joined the movement. William E. Akin, in *Technocracy and the American Dream,* states that one of the sources of technocracy was "the search of the engineering profession for an occupational identity." It is true that some engineers, resentful of being dominated by businessmen and enchanted with the idea of applying rational methods to social problems, were thinking in terms of new leadership roles for their profession. But as Edwin T. Layton, Jr., has noted in *The Revolt of the Engineers,* technocracy "represented a grotesque parody of the engineers' thought, rather than a legitimate expression of it." Worse, the movement "tended to discredit the profession's reform tradition."

The idea that society can be, or should be, "engineered" appealed only fleetingly to a handful of engineers, and then was discarded as being impractical, if not immoral. But the concept appealed to the general public in a time of crisis, and that is what makes the technocratic craze so interesting. The feeling of helplessness experienced during the Depression brought to mind the efficient operation of control boards during World War I, and started people thinking about putting experts in charge of things. A trace of the same exasperation exists today. "If we can put a man on the moon, why can't we. . . ?"

The technocratic concept can be seen to be a jumble of absurd inconsistencies. It starts with intellectuals who want to enlist science in the cause of social justice. It derives emotional intensity from a naive public faith in technical experts. But because science can neither establish standards of virtue nor encompass the vagaries of human desire, all technocratic schemes must inevitably fail; indeed, they must inevitably appear ridiculous.

Where the idea continues to be proposed seriously, it descends to the level of caricature. Take, for example, Ayn Rand's novel *Atlas Shrugged,* in which the engineer–hero, John Gault, frustrated by the activities of bureaucratic do-gooders, leads his fellow professionals in a withdrawal from society. Trains stop running and factories grind to a halt. Eventually the engineers are invited to return upon their own terms, totally free from regulation by nontechnical politicians. Ayn Rand shows herself to be an extravagant ideologue, a right-wing descendant of Henri de Saint-Simon and Howard Scott. Needless to say, John Gault is unlike any real-life engineer one is liable to encounter.

With technocracy discredited, a strange mutation has occurred. The word "technocrat" has been converted from a term of hopeful praise into an epithet. The scientist-ruler appears in fiction as Dr. Frankenstein, or Dr. Strangelove, or even James Bond's arch-enemy, Dr. No. Engineers are accused of using their special skills in the sinister exercise of power. Thus have intellectuals, disappointed in their quest for technocratic order, vented their frustration in unwarranted attacks upon technically trained professionals.

# 4

# The Myth
# of the
# Technocratic
# Elite

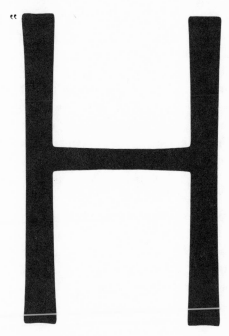

ere then are our new masters," proclaims a Harvard professor of government in *The New York Review of Books*—"bureaucrats, technocrats, scientists, and their professional allies . . . people with specialized training and knowledge." [9]   ¶The idea is false, yet it is widely expounded. The myth of the technocratic elite continues to grow, fostered not only by the complaints of disgruntled humanists, but also by the theories of prestigious social scientists. Engineers, who daily face their political impotence, are amazed to hear that they have become a ruling class.

Perhaps the early proponents of technocracy, from Saint-Simon to Veblen, did hope that engineers and scientists would take over the leadership of society, but they never went so far as to claim that it was actually happening. Thus Veblen in 1919:

> Popular sentiment in this country will not tolerate the assumption of responsibility by the technicians, who are in the popular apprehension conceived to be a somewhat fantastic brotherhood of over-specialized cranks, not to be trusted out of sight except under the restraining hand of safe and sane businessmen. Nor are the technicians themselves in the habit of taking a greatly different view of their own case.[10]

The technocratic "movement" of 1932 was, as we have seen, a fiasco that, by all logic, should have permanently discredited the notion of technocracy. But with the growth of technology during and following World War II, the idea was revived.

The coming of the technocratic age was announced in 1941 with the publication of James Burnham's book *The Managerial Revolution.* According to Burnham, a social revolution was underway, but instead of the proletariat supplanting the capitalists, there was arising a new ruling group called "the managers"—more specifically the "production managers," "administrative engineers," and "supervisory technicians."

Nothing of the sort was really happening. Nevertheless, as American society entered, in rapid succession, the nuclear age, the space age, and the electronic age, Burnham's thesis gained credence. Occasionally it was lent support by the comments of irate politicians such as when President Eisenhower, in his farewell address warned about the growing influence of a "scientific–technological elite."

When C. Wright Mills wrote *The Power Elite,* published in 1956, it was not his intention to lend support to the technocratic myth. In fact, Mills believed that power belonged to "the warlords, the corporation chieftains, and the political directorate," and that the learned professional was usually reduced to the role of "a hired technician." But, in arguing that power was vested in those who have "access to the command of major institutions," Mills seemed to be endorsing the technocratic concept. Perhaps he should have used a term like *the power grabbers,* which is closer to what he meant than *power elite,* a term that, if it means anything, implies the relationship of knowledge and skill to the management of human affairs.

It was only a matter of time before someone would write a book entitled *The Technical Elite,* and this was done in 1966 by Jay M. Gould, an economist who had conducted statistical readership surveys for *Scientific American.* Mr. Gould's figures showed that the number of engineers and scientists was growing rapidly and that many of them had well-paid jobs in industry. From this he deduced that a powerful new class of technically trained people was taking shape. I have often wondered what Mr. Gould had to say when, in the early 1970s, cutbacks in the aerospace industry led to the wholesale firing of thousands of the members of this invincible class.

At about the same time that Gould was collecting his statistics, Don K. Price, dean of the Graduate School of Public Administration at Harvard, was writing *The Scientific Estate.* Price saw that the growth of science and technology was accompanied by an exceedingly complex pluralistic ferment, and he spoke of the emergence of four "estates"—the pure scientists, the professionals, the managers, and the politicians—each with its own strengths and limitations. Price discerned the evolution of a new system of checks and balances: politicians and bureaucrats are restrained from taking action that would fly in the face of scientific reason and bring down on their heads the scorn of the scientific community. On the other hand, scientists and engineers are unable to attain real power except as they might manage to get it through established political channels. This seemed to me at the time, as it still does, an eminently sensible way of looking at the evolving social structure. The public, however, impatient with complexity, and perhaps titillated by residual dreams of royalty, seemed to prefer theories that purported to identify ruling cliques.

All during the 1960s there was a veritable blizzard of books dealing with the ever fascinating question of power. Just the titles quicken the imagination: *The American Establishment, The Super-Americans, The Power Structure, Wealth and Power in America, The Higher Circles,* and myriad others. Although the authors of these books were more often impressed by wealth and social connections than they were by technical training, the persistent reference to such terms as *elite* and *establishment* encouraged people to think in a technocratic mode. If we do indeed live in a society dominated by a select caste, if the membership of this caste is changing, and if American society is technological, then it stands to reason that technologists will somehow gravitate toward positions of power.

The concept of an existing technocracy took a quantum leap in public recognition with the appearance in 1967 of John Kenneth Galbraith's *The New Industrial State.* "Effective power of decision," wrote Galbraith, "is lodged deeply in the technical planning and other specialized staff." The people ordinarily associated with the exercise of power, in government as well as in industry, are merely "ratifying" decisions made by the technicians. This is an interesting idea, and doubtless there are instances when events are determined by the invisible counselors to the mighty. But, as G. William Domhoff has written in *Who Rules America?,* "To advise a decision-maker is not to make a decision. . . . It is the function of the decision-maker to choose among the usually conflicting advice that he receives from his usually divided experts." Nevertheless, Galbraith's book did wonders for the image of the technologist as a wielder of power.

Daniel Bell's *The Coming of Post-Industrial Society* (1973) was written in the same vein. "In the post-industrial society," Bell wrote, "technical skill becomes the base of and education the mode of access to power." Although he followed this statement with a goodly number of qualifications, and eventually with the conclusion that "real power" is political rather than technical, the technocratic implications of Bell's work were what left an impression.

With such respected social scientists as Galbraith and Bell providing a theoretical underpinning, and with the effects of technology becoming daily more evident, the existence of a technocratic elite has come to be increasingly accepted as a reality. This is happening in spite of evidence that indicates that technocracy is as much a fantasy as ever. Indeed, if one is forced to generalize, it is more accurate to characterize engineers as an exploited class than as "our new masters."

Of the college degrees awarded in the United States in 1948, 11 percent went to engineers. Since that time, although the number of engineering degrees awarded annually has nearly doubled, the total number of college degrees has quadrupled, so that at the present time engineers constitute only about 5 percent of each year's degree earners. (These figures pertain to master's degrees, as well as bachelor's. At the doctoral level, the figure was 6 percent in 1948, rose to 13 percent in the mid-1960s, and then gradually declined to 8 percent.) This decrease in the percentage of engineering students is largely a result of the growing numbers of women attending college, very few of whom have entered engineering.

Cumulatively over the past 30 years, somewhat less than 7 percent of all Americans graduating from college have been engineering majors. However, engineers constitute 9.5 percent of all employed, college-educated, nonclerical, white-collar workers. Again the difference between these two percentages can be explained largely by the activities of women, many of whom have graduated from college but not remained in the workforce.

The significance of these figures is twofold. First, since the percentage of engineers among college-educated Americans has been dropping for 30 years, numerically at least, they are not "taking over." Second, if we live in a technocratic society, we would expect to find engineers in many positions of power, certainly in greater proportion than the 9.5 percent they constitute of the college-educated workforce. But this is not the case. Wherever there is power it will be found that engineers are underrepresented.

Take, for example, *Who's Who in America.* According to one school of sociological thought, America's ruling class is alive and well, and its members can be identified by looking through social registers, membership lists of select clubs, interlocking corporate boards, and so forth. *Who's Who* serves as a distillate of such standards of eminence, so it can be said to provide a guide to the makeup of the ruling class. If one takes the trouble to classify by educational background the 72,000 people listed in *Who's Who in America,* it will be found that only 4 percent are engineers by training, and many of these are academics, unarguably distinguished but hardly powerful.

Perhaps it is no great surprise that engineers are not to be found in patrician circles. However, one would certainly expect to find them abundantly represented in the high levels of government bureaucracy. An analysis of *Who's Who in Government* seems at first to indicate that in this area they are at least holding their own. Approximately 9 percent of the people listed in this publication—senior officials at the federal, state, and local levels—are graduate engineers. But of the 40,000-odd individuals in the federal government's most senior civil service positions—GS grades 15–18 (roughly equivalent to Major through General)—the representation is as follows:

| Grade | Engineers |
|-------|-----------|
| 15 | 16% |
| 16 | 9% |
| 17 | 6% |
| 18 | 5% |

At the top of the bureaucratic pyramid the engineer is jostled aside.

Outside of the civil service the situation for engineers in government is even worse, with their presence dwindling to practically nothing. Of the 6,000 professionals employed on Capitol Hill, fewer than 1 percent are engineers. According to a mid-1980 report, about 13 engineers work for the House Science and Technology Committee, two work on the House Subcommittee on Energy and Power, and about a dozen for the Office of Technology Assessment.[11] Very few Congressmen employ engineers on their staffs. As for the Congress itself, in mid-1980 the newsletter of the American Society of Engineering Education published an interview with Donald Ritter: "The Only Engineer in the United States Congress." The society had to publish an apology when, among the 535 members of Congress, a second engineer was discovered.

In the executive branch of the federal government there is a similar paucity of engineers at or near the top. Of course, Jimmy Carter studied engineering at Annapolis, but he spent most of his pre-political years as a farmer–businessman. George Washington was, among many other things, a land surveyor–an oddity that is emphasized each year by the National Society of Professional Engineers when they celebrate Engineers Week around February 22. But the only real engineer–President was Herbert Hoover, a worthy and much-maligned man who nevertheless managed to tarnish indelibly the image of the technologist as political leader. Before becoming President, Hoover served as Secretary of Commerce, and since his time there have been precious few technologists at, or even near, cabinet level.

The most famous "technocrats" of recent times have been falsely labeled, at least if one pays any attention to the root meaning of words. None of them has been a trained technologist, that is, an engineer. James Schlesinger, denounced as a technocrat throughout his years as Chairman of the Atomic Energy Commission, Director of the Central Intelligence Agency, Secretary of Defense, and Secretary of the Department of Energy, received a Ph.D. in economics from Harvard. He taught that subject for eight years at the University of Virginia before entering government service as Assistant Director of the Bureau of the Budget. Robert McNamara, who purportedly started a technocratic crusade during his term as Secretary of Defense in the early 1960s (Daniel Bell claims that "the McNamara 'revolution' represented a rationalization of governmental structure") earned an M.B.A. at Harvard Business School, taught business administration at Harvard, and rose to be president of the Ford Motor Company via the post of controller.

George Romney, who, as Secretary of Housing and Urban Development (1969-1972) proposed to mechanize the nation's construction industry, was a college drop-out and salesman before he worked his way up to become president of American Motors.

Only if the term *technocracy* is expanded to signify rule by economists, business managers, lawyers, and accountants, as well as by scientists and engineers, can it be suggested that we are entering a technocratic age. But this stretches the word beyond all reason. To observe that clever, well-organized people, using the most up-to-date organizational methods, rise high in government is to iterate the obvious. By these standards, it could be said that Richelieu, Talleyrand, Metternich, Hamilton, and a thousand other political leaders of the past were technocrats.

Even if the meaning of technocrat is thus extravagantly expanded—using "technique" as the root instead of "technology"—the place of technocracy in our society is far from being established. When one considers how such haughty pseudo-technocrats as James Schlesinger can be frustrated by Congress, hounded by industry, tormented by the press, and peremptorily discharged by the President, the role of the "expert," no matter what he is called, can be seen to be far from dominant.

Speaking of discharging experts, one cannot help but be reminded of the sad history of the President's science adviser. The position was created in 1957 in the aftershock of Russia's Sputnik triumph. Backed by a full-time staff, and assisted by a prestigious President's Science Advisory Committee, which met every two months, the adviser was charged with providing independent counsel on technological issues to the President and the Budget Bureau. Although the post was held by a series of eminent and respected individuals, starting with James R. Killian, Jr., former president of MIT, and ending with Edward E. David, Jr., of Bell Telephone Laboratories, the committee repeatedly found itself subject to manipulation and misuse. Its recommendations concerning DDT were received but not implemented. Its warnings about underground nuclear weapons testing were suppressed. Its report on defoliation, superficial to begin with, was not released until after the program in Vietnam had been ended. In the SST and ABM debates, the committee's warnings were hidden behind a barrier of "confidentiality," while its endorsements were used to support the administration's arguments. When individual committee members made remarks in public that President Nixon considered disloyal, the entire organization was summarily disbanded.

In 1976, during the Ford administration, Congress restored the President's science adviser, and put him in charge of a new Office of Science and Technology Policy (OSTP). But the founding legislation carefully limited OSTP in size of staff and scope of responsibility and failed to restore the President's Science Advisory Committee.

Another example of technologists subservient to politicians is to be found in the recent history of the National Academy of Sciences. The academy was established by act of Congress in 1863, but is essentially a self-governing body of considerable repute (not to be confused with the National Science Foundation, a government agency established in 1950 to funnel federal funds into promising research and development programs). This noble institution is often called upon to serve as an advisory body for the government, and in theory its recommendations carry weight. Yet in studies of radioactive waste disposal, the SST, defoliation, food additives, pesticides, and airborne lead, the academy's Research Council has shown a disconcerting tendency to report its findings in ways least likely to offend whichever administration had commissioned its work. Consequently, the academy has suffered embarrassments comparable to those of the President's advisers. In relations between client and professional, it is inevitably the client who has the final say.

There are other nations in which the situation is different. In France, for example, the governing elite contains many graduates of the Ecole Polytechnique, a highly respected technical university that dates from the eighteenth century. (This is in stark contrast to the English tradition of an "old boy" network of liberally educated aristocrats.) In Russia most members of the Politburo have been educated as engineers. (Not that this seems to have perceptibly changed the quality of life. According to the dissident scholar Roy A. Medvedev, since the world is becoming increasingly complex, "Scientists and specialists are being drawn into the *apparat* ... but at the same time undemocratic methods of administration remain intact." [12])

In the Third World engineers are frequently found in responsible government positions. In developing societies technologists are both rare and esteemed. Also, in such nations engineers often come from well-to-do families and are educated in Western schools where they are exposed to progressive political thinking. Whatever the reasons, there they are: Moustafa Kahlil, premier of Egypt, Mehdi Bazargan, the first prime minister of the Islamic Republic of Iran, Yassar Arafat, chairman of the Palestine Liberation Organization, José Napoléon Duarte, president of the ruling junta in El Salvador, and many others.

But other nations are other nations. In the United States the lowly role in government affairs assigned to—and accepted by—

scientists and engineers makes the incessant references to technocracy seem anachronistic.

Of course, in the United States, one can be excluded from the moneyed aristocracy and from the new social elite, be denied a role in government, and barred also from such centers of influence as the media, the arts, the banks, and the large law firms, without necessarily being deprived of power. There remains for the engineer a position in what many people would call the true center of power in our society—American industry.

Certainly engineers are numerous in industry, and they rise to positions of responsibility. *Who's Who in Finance and Industry* lists approximately 18,000 individuals in senior management, of whom about 26 percent are engineers by training. But as is the case in the federal civil service, a smaller number make it to the very top. In the country's 100 largest corporations during the late 1970s, slightly less than 15 percent of the chief executive officers had engineering backgrounds. This was up from the 9 percent indicated by a 1968–1972 study, but down from an 18.5 percent high of 1958–1962.

Most chief executive officers, according to the latest survey, are individuals with financial backgrounds (23 percent), with marketing people (19 percent) slightly behind.[13] (To say that engineers occupy 15 percent of the top places in industry is not to say that they occupy 15 percent of the top places in the commercial world. When leading bankers, lawyers, insurance executives, and other specialists are included in the figures, the representation of engineers in the highest echelon of money-makers is reduced to less than the 9.5 percent they represent of the college-educated workforce.)

Beyond such bare statistics lie many perplexing questions, a number of which are addressed in a 1979 study conducted by Korn/Ferry International, a large executive search firm, assisted by a group from the UCLA Graduate School of Management.[14] The study covered more than 1,700 senior-level executives in the leading industrial corporations (The "Fortune 500"), and the largest banks, insurance, retailing, and transportation companies (the "Fortune 50's"). Of the people constituting the "executive profile," 31 percent had started their careers in professional/technical jobs. But when asked what they considered most important in bringing about success, only 2.5 percent of the group mentioned "professional or technical competence." This item was twelfth on a list headed by such factors as "hard work" (17.5 percent), "ambition" (9.5 percent), "human relations" (9 percent), and "timing" (4.5 percent). Less than 7 percent of the sample thought that the "fastest route to the top" lies in the professional/technical area; most of them voted for finance/accounting, marketing/sales, or general management. Even when asked about the *future* "fastest route to the top," the choices changed little, with the professional/technical selection rising only to 10 percent.

This is in marked contrast to what appears to be happening in Japanese industry. According to a 1980 *Wall Street Journal* dispatch from Tokyo, "More than a dozen key companies in the past year or so have chosen presidents with backgrounds in engineering. Industry analysts say this trend away from the usual expertise in sales and finance reflects the attempt of Japanese companies to gear up for what they see as the need for the 1980s." [15]

But even in the Japan of the 1980s—and beyond—it is doubtful that technocrats will achieve dominance. Already, according to the *Wall Street Journal,* there are complaints about the engineer executives: they tend to be "too rational," and even worse, they aren't "fun-loving." There has been some dismay expressed over the fact that the engineer–president of Toshiba dislikes that traditional part of Japanese business negotiations, the geisha party.

No matter how complex technology becomes, and no matter how important it turns out to be in human affairs, we are not likely to see authority vested in a class of technocrats. To be sure, the well-educated citizen will be increasingly knowledgeable about technology, and engineers will continue to fill many important roles. But the social forces that keep engineers out of *Who's Who,* and prevent them from reaching the pinnacles of government bureaucracy, will continue to operate. Even in industry it can be assumed that "hard work" and "ambition" will continue to be the keys to executive status, along with "human relations" and serendipitous "timing." It is no accident that most engineers who reach positions of influence do so by dint of becoming more social, inevitably at a sacrifice of technical proficiency. As described by a retired executive of the Ford Motor Company, "Engineering managers are chosen on their ability as managers, how they meet their schedules and budgets, how they manage their people, how they control their costs. . . . As a consequence, the engineering management jobs are highly likely to go to people who have inferior ability and judgment in technical matters." [16] Not only are engineers underrepresented in the ruling classes, but those who are there have generally given up the practice of engineering.

The myth of the technocratic elite is an expression of fear, like a fairy tale about ogres. It springs from an understandable apprehension, but since it has no basis in reality, it has no place in serious discourse.

# 5

# Hired Scapegoats

n 1971, about a year after the first celebration of Earth Day, General Frank Koisch, Director of Civil Works, U. S. Army Corps of Engineers, was invited to address a meeting of college newspaper editors in Washington. His announced topic was "Does the Corps Give a Dam?" A more immediate question was, Would the general be allowed to speak? From the moment Koisch strode into the hall, resplendent in military uniform, the audience of young journalists was in an uproar. For almost half an hour a battle raged between the program committee, pleading for silence, and a group of hostile activists, determined that this enemy of the people should not be heard. It was finally agreed that, no matter how repugnant his views, an invited guest should be permitted to speak. Order was restored, and General Koisch stated his case, which was that the corps did give a damn.    ¶There is no evidence that he convinced anybody present. As he spoke, a young woman circulated through the audience handing

out bumper stickers that read "Dam the Corps—Not Our Rivers."
When the general had concluded his remarks, it was discovered
that his hat had been pilfered and that someone had carved
obscenities on his leather briefcase.

The incident was a sign of the changing times. The heroic
image the U. S. Army Corps of Engineers had enjoyed in an earlier
age vanished in the aftermath of Vietnam and the environmental
crisis. Its dam-building, dredging, draining, and other works,
which once seemed so marvelous, are regarded increasingly with
revulsion. Although never without its few vocal critics (Harold L.
Ickes complained that it was "above the law"; Justice William O.
Douglas labeled it "public enemy number one"), the corps could
not have been prepared for the virulent hostility directed against it
during the 1970s. Books with such titles as *The River Killers* and
*Dams and Other Disasters* chronicled the corps' alleged predations
and called for its dissolution. Magazine articles—from "Dam Outrage" *(The Atlantic,* April 1970) to "Flooding America in Order to
Save It" *(New Times,* November 1976)—characterized it as "a giant
bulldozer out of control, burying villages, disfiguring the landscape." The other media contributed to the swelling expression of
public outrage.

Even politicians, who once treated the corps with deference,
joined the attack. Congressman Stewart Udall compared the corps
to "a giant water-loving dinosaur with less brain per pound of flesh
than any other vertebrate." Senator William Proxmire awarded
the corps his 1976 "Golden Fleece of the Year" for "the worst record of mismanagement and cost growth in the entire government."

Upon reflection, there is nothing at all surprising about this
development. A more fitting villain for this nation in this era
could hardly be imagined. The U. S. Army Corps of Engineers
appears to embrace in one entity the three segments of American
society that evoke our most intense protest: the military, the bureaucracy, and the environment-savaging technocracy.

But appearances, as we continually say and repeatedly forget, can be deceptive. In the case of the Corps of Engineers, the publicly accepted image happens to be completely at variance with the facts. This does not move me to mount a defense on behalf of the corps, which has, after all, not suffered anything more cruel than a bad press. But I think the matter deserves examination because it exemplifies a combination of public misunderstanding and frenzy that seems to be a recurrent feature of our national behavior. It also demonstrates that even where engineers are nominally in control of important enterprises, their work is responsive to the needs and desires of the community in which they live.

What are the facts about the Corps of Engineers, and how do they differ from the image?

That part of the corps which builds dams, dredges harbors, and attends to other civic works—the Civil Works Directorate—is simply not, by any reasonable definition, a part of the military establishment. Although technically a branch of the Army, this organization is, in fact, an agency of the United States government. The directorate's 40 offices across the nation are staffed by 32,000 civilian engineers, technicians, and other civil servants. A mere 300 Army officers nominally oversee the activities of this huge organization, and their involvement is circumscribed by the fact that their service in the directorate is limited to three-year tours of duty. More important, the power to authorize the study of a corps project, initiate it, and appropriate the money for it is held, not by any arm of the military, but by the Public Works Committees of the Congress and the Public Works Subcommittees of the Appropriations Committees. The Secretary of the Army rarely interferes in these matters. Even the Budget Bureau and the White House think twice before getting involved. The Corps of Engineers is an agency through which Congress studies, evaluates, and executes public works projects, particularly in the area of water resources development.

Why, then, the anachronism of keeping this institution as a branch of the Army? Why not establish it as a government agency known simply as the Department of Engineering, into which other federal engineering organizations could also be integrated, including the Bureau of Reclamation of the Department of the Interior, which provides irrigation facilities for the 17 Western states? Someday it may come to pass. Common sense favors the idea; tradition, however, opposes it.

Engineering, for almost all of recorded history, was closely linked to the military. Fortifications and weapons were major engineering concerns. Transport and water supply came within the province of military planning. The term *civil engineer* did not even exist until the mid-eighteenth century, when it was coined by the famous English engineer John Smeaton, builder of the Eddystone lighthouse, in an attempt to differentiate his work from that of the military. The United States Military Academy at West Point was established in 1802 as an engineering school, and for several decades was the main source of engineers for the nation. When Congress embarked on mapping the unexplored West, and developing harbors, canals, and other massive public works, it quite naturally got into the habit of delegating projects to the Army Corps of Engineers. A tradition so entwined with the history of our land is not quickly cast aside. For all practical purposes, however, the corps has almost nothing to do with the military.

What of the corps' reputation as an arrogant, unresponsive bureaucracy? Here again the facts belie the myth. The *sine qua non* for an unresponsive bureaucracy is an established, independent, and relatively invulnerable fiefdom. The Civil Works Directorate has nothing of the sort. Each year the Public Works Appropriation Act provides funds for the corps' civil-works program on a project-by-project basis. No other major federal agency has its work funded in this way. Critics of the corps say that the annual appropriation for each project serves to obscure the long-term cost of these projects. At the same time, however, it also makes corps activities supremely sensitive to every wish of Congress. The appropriation of funds for the corps is, in fact, the major "pork-barrel" legislation of each Congressional session, and it reflects unerringly the mood and the shifting power relationships in the Senate and House. Projects are started, stopped, expedited, and delayed, and the action is parcelled out in each of the 50 states according to agreements arrived at in the labyrinths of the Capitol.

Doubtless some members of the corps have learned their way about in those labyrinths and have proved themselves adept at such bureaucratic tricks as juggling cost-benefit ratios and rationalizing tremendous cost overruns. But it is clear that they have no real power, being dependent at 12-month intervals on Congressional whim. Far from being an intransigent bureaucracy, the corps appears to have evolved as an instrument exquisitely tuned to work the will of the people.

All right, critics of the corps might concede, but which people? Corps projects traditionally come into being when some local citizens' group gains the political support of a Congressman and the technical approval of the local corps district engineer. Typically, the local group is a Chamber of Commerce or some other representative of monied interests. Yet even if many projects are conceived in greed, and sponsored under slightly unsavory circumstances, the entire local community often benefits from increased employment and a prospering business climate. Some studies have sought to demonstrate that other types of federal programs would be more effective in aiding local communities, and that is probably true. But pending the evolution of such programs, the Corps of Engineers serves as a conduit for Congressional revenue-sharing. The gains of sponsoring entrepreneurs, ill-gotten as they may be, are not in the long run a major consideration. As A. Den Doolard, a Dutch author, has written of the contractors who work on building the dikes in Holland: "Profit is merely the bait that destiny has offered to these calculators."

The destiny of America, as perceived until recently by the vast majority of its people, has been to grow economically and to develop its water resources to this end. Wilderness areas have been flooded, rural families uprooted, archeological sites inundated, and important caves damaged, not because these were objectives of the Corps of Engineers, but because commercial development was mandated by the citizenry. As the values of people change, and as Congress reflects such change, the activities of the corps will change automatically. Indeed, changes are occurring at this time. Two college professors, writing in *Public Administration Review,* claim that such changes demonstrate how aging, entrenched bureaucracies are capable of innovative and progressive response to new conditions. But this misses the point about the corps, which is that, because of the unique year-by-year, project-by-project funding of its activities, it is not a comfortable, unresponsive bureaucracy.

If the corps is neither a true branch of the military nor an entrenched bureaucracy, does it not at least stand condemned for its technocratic destruction of the environment? The issue is complex, but again the corps must be found not guilty.

If, by doing away with our inland waterway system, we could have back our wild rivers, how would we then transport the one-sixth of all intercity cargo that is presently water-borne? By truck and railroad, of course. Comparing barges moving slowly upstream with roaring trailer trucks and freight trains, I, for one, cannot see where this would be much of an environmental improvement. The corps has been accused of being in cahoots with the barge industry; this may be true, but since such complaints usually come from railroad and trucking lobbyists, they do not excite the environmentalist in me.

Another corps activity that has troubled environmentalists is in the area of water supply. During those years in which we have been blessed with adequate rain and snow, the people who see fit to excoriate the builders of dams and reservoirs have not had to worry about drought. But as soon as precipitation drops below acceptable limits, we are haunted by visions of failed crops and incipient dust bowls. Water shortages are "natural" occurrences, I suppose, but some modification of the environment in an effort to avert such disasters seems to me to be morally justified, even in a society sensitive to ecological concerns. Surely the majority of our citizens have supported and continue to support public works projects whose purpose is to assure an adequate supply of water.

On the other hand, in the area of flood control, things have been done which are difficult to defend. By damming and leveeing, and permitting commercial and residential development of the flood plains, the corps has restricted rivers to artificial channels, where they flow more swiftly and become potentially more dangerous than they ever were. Now belatedly, the corps is stressing "non-structural" methods of flood control. It is only proper that it be held accountable for its past errors in this field. But in a nation where people persist in living on cliffs which are crumbling into the sea, and build houses atop major earthquake faults, there is little likelihood that even the most far-seeing engineers could have prevented the rush onto the low-lying plains. It is a lot easier to hold back torrents of water than it is to stand in the way of land-hungry Americans.

Perhaps the most serious problems caused by the corps result from its indiscriminate filling, dredging, and draining of our wetlands. But it is hard to place the blame for these acts entirely, or even mainly, on corps personnel. Until recently the importance of wetlands in the ecological scheme of things was not understood. Estuaries and swamps, we have only lately learned, moderate our climate, provide natural pollution control, and play a vital role in the life-cycle of a multitude of marine organisms and other animal life. If these facts were not sufficiently known to biologists, meteorologists, agronomists, and other environmental specialists in our society, how can we expect the corps' civil engineers to have been uniquely prescient? Considering the general lack of knowledge in these matters, was it such an inexcusable manifestation of hubris for the corps to have tinkered with nature? Hardly—unless we condemn all humanity for its dreams, dating from earliest times, of draining malarial swamps and "reclaiming" what appeared to be fetid marshland. Goethe's Faust, remember, found his final salvation in a land-reclamation project.

In light of the new scientific knowledge, Congress in 1972 gave the corps responsibility for protecting all the wetlands in the nation. When the corps defined its responsibility as limited to its traditional province, the *navigable* waters, environmental groups brought suit to establish the corps' authority over *all* waters. In a situation not without irony, the courts agreed with the environmentalists that the corps should have total control. Taking its new responsibility with all earnestness, the corps in 1976 stopped the Deltona Corporation of Florida from turning 2,000 acres of mangrove swamp into a housing development, even though Deltona had already sold the land to prospective home-builders. Commercial developers all over the nation were aghast. "The decision was a shock," said the president of Deltona. "I still can't get over it. The corps—they've been like us. They're engineers, our kind of people."

Obviously, the president of the Deltona Corporation, like a lot of people in this country, has no understanding of what engineering is all about. The profession is dedicated to performing works "for the general benefit of mankind" or "for the good of humanity," to quote from two definitions of long standing. But this does not mean that engineers take unto themselves the right to define what such benefit or good might be, or, even less, that they are committed to a real estate developer's ideas on the subject. Society establishes its own goals, and engineers, like jurists, educators, politicians, and the rest of the body social, work toward achieving these goals. When the nation wanted to fill in its wetlands and tame its rivers for the sake of commerce, the engineer did the job. If the nation has become more sensitive to environmental considerations, then so, by definition, has the engineer. Engineering is not anti-environment. Environmentalism itself is a branch of engineering.

This is not to say that engineers are automatons without

conscience or conviction; they are philosophically an integral part of the community. This is a thorny issue to which we will have to return in a discussion of engineering ethics in chapter 15. Engineers are called upon to be guided by conscience at the same time that they are urged to serve the will of the public. As far as the U. S. Army Corps of Engineers is concerned, there is ample evidence to demonstrate that it is responsive to the desires of "the people."

Members of the corps are action-oriented, to be sure, but they are no more devoted to building dams than they are to protecting wetlands. In fact, in the Mississippi delta they are experimenting with ways of creating *new* wetlands. A certain amount of foot-dragging is inevitable among those long associated with particular undertakings. But the brightest, most alert, and most ambitious members of the corps will see to it that the times do not pass them by. Careers are not made by defying the will of the electorate.

Some of the corps' new ecological awareness will be discounted as cynical lip service. When the chief of engineers jovially passes out big buttons saying "The Corps Cares," this proves only that a public relations department is at work. But wanting a reputation for caring can clearly be considered a step in the right direction.

In response to the changing national mood, the corps is constantly reevaluating and deauthorizing many of its projects; at the same time, it is moving ahead with some projects to which there is much public opposition. Even after new standards of caution and sensitivity are applied, there remain areas of disagreement which are essentially a matter of taste. What sort of a landscape do we want? What mix of wilderness and factories, parks and highways, suburbs and cities *will do?* It is said that there is no disputing taste, but in fact there is nothing more important to dispute. John Dewey put it this way:

> The formation of a cultivated and effectively operative good judgment or taste with respect to what is esthetically admira-

ble, intellectually acceptable and morally approvable is the
supreme task set to human beings by the incidents of experience.[17]

This challenge is one that the citizens of a free society must
recognize and accept. We cannot avoid it by pretending that our
fate is in the hands of organizations such as the Corps of Engineers.

There is no way in which we can recapture the wild conti-
nent that once was. To regret this, I believe, is an elitist conceit.
But we can stop any new project at any time. All we have to do is
convince ourselves, and then our Congress, that this is what we
want to do. The Carter administration, when it first came into
office, selected 59 water-resource development projects as "high-
priority projects for reevaluation" and then recommended a halt to
19 of them. It discovered that objections came, not from engineers, but from ordinary citizens and their Congressmen, who
chose continuing commercial development of their home areas
over environmental concerns and budgetary restraint. The experience of the Reagan administration was similar.

We say that we oppose the corps for being militaristic, bureaucratic, and anti-environmental, but upon inspection these reasons are seen to be invalid or feeble. We actually oppose the corps
because it so unerringly shows us what we are—or what we just
were.

The concept of scapegoat has come down to us from biblical
myth. Perhaps, like so many mysterious phenomena of mass psychology, a combination of confused and misdirected blame is, in
some way that we cannot see, a vital element in maintaining social
stability. This must be the hope of those who wish not to be overly
discouraged by recent public behavior toward the U. S. Army
Corps of Engineers.

# 6

# Nuclear Angst

*uid erat demonstratum.* There is no technological imperative. There is no reigning technocratic elite. My mode is conversational, and I have offered no formal proofs. But I believe that the facts speak for themselves. Yet a feeling of uneasiness persists. Are we not doing certain things against our better judgment? Are we not moving relentlessly along treacherous paths, staggering toward disaster like a mob blind or possessed? No normal person is immune to such anxieties. For me, as for many others, there is one technology that is uniquely able to trigger an uneasy reaction, and that is the technology of nuclear power.

Nuclear weapons, of course, are the greatest technological threat, and they are doubtless responsible for much antitechnological dread. Yet it will not do to blame the horrors of warfare—even the unthinkable horrors of nuclear warfare—on technology. Too many millions of innocents have been butchered by the sword—indeed, by the club—for us to do that. In any event, all sane people, technologists and antitechnologists alike, will certainly exert every effort toward seeing that nuclear weapons are never used, and I must let it go at that, for the subject of world peace is beyond the scope of this essay.

But the subject of nuclear power cannot be sidestepped, not if I am to pursue the argument I have commenced. Atomic energy is part of our daily lives, and daily by our action—or inaction—we give our answer to the charge that it is developing independently of the purposes of society. It is an example that tests to the limit, but in the end I think supports, my contention that technology is an expression of the aggregate communal will.

Connecticut Yankee, the atomic power plant that has produced more electricity than any other nuclear generating unit in the world, is located on a 500-acre wooded site on the east bank of the Connecticut River in the town of Haddam, Connecticut, southeast of Middletown. It is surrounded by some of the most picturesque scenery in the American northeast. Steel truss bridges, telephone poles, and red brick warehouses blend comfortably into the landscape, along with white steeples and brightly painted antique shops. In the New England countryside, technology is at its most benign.

As I drove toward the plant one morning in early January 1979, there was no indication that I was approaching a large power-generating facility–no smoke, no noise, no trace of heavy commercial traffic–and I thought of how, just a few years earlier, nuclear power was considered to be a dream come true, a clean, cheap, compact source of energy that would carry our civilization to new levels of sparkling achievement. In December 1962, when a group of New England utility companies formed the Connecticut Yankee Atomic Power Company, and planned to build a nuclear plant at Haddam, a mood of buoyant anticipation prevailed. There were no "intervenors" to challenge the project with extended lawsuits, no Clamshell Alliance to stage demonstrations on the proposed site. Ralph Nader was an unknown lawyer just about to take up the cause of automobile safety. In mid-1963 the Atomic Energy Commission approved an application for construction of the new facility, which was then designed, reviewed, and built at a pace we can only marvel at today. When commercial operation began on January 1, 1968, just five years had elapsed from the time of the initial application to AEC. The total cost of the project amounted to a little more than $100 million. At the time of my visit, the lead period for a nuclear plant had become ten to twelve years, and the cost was approaching $2 billion.

For Connecticut Yankee the first 11 years had been an almost uninterrupted success story: more than 45 billion kilowatt hours produced at one of the nation's lowest per-kilowatt costs. For the nuclear industry as a whole, however, those same years had seen great expectations dashed, as reactor orders, which reached a high of 41 in 1973 dropped to four in 1975, then three, two, and two again, recorded in the final days of 1978. Despite the success of Connecticut Yankee and the 71 other atomic fission plants that were producing 13 percent of the nation's electricity, the nuclear power industry appeared to be dying.

Incredibly, this was happening at the very moment that the political and economic costs of oil were proving unbearable, the hopes for doubling the use of coal were seen to be unrealistic (and environmentally horrendous), and the near-term contributions of alternative energy sources were known to be minimal.

In the nuclear power debate, however, facts seem to lose their force, and incongruities abound. Why, for example, do left-wing activists choose to march on nuclear power plants when it is the poor who suffer most from the escalating cost of energy? And why should being a conservative make one inured to fears—even irrational fears—about the health of one's children? The anti-nukes wear sneakers and are called elitists; the pro-nuclear forces wear vested pin-stripes and preach the needs of developing nations.

I was on assignment from *Harper's,* and I was determined to succeed, where so many others had failed, in making calm sense out of the bewildering nuclear power debate. I did not know, of course, that the accident at Three Mile Island lay just 12 weeks in the future.

At the plant I was met by Tony, a jovial ex-high-school science teacher who presided over what is called the Energy Information Center, a small museum open to the public. "It's not a hard sell," Tony said as he showed me through the exhibits. And indeed it was not, the colorful displays explaining electricity in general terms, with each type of energy, including wind and solar, getting a fair and factual demonstration. Nuclear fission, which was explained with the aid of an animated cartoon, seemed as harmless as the whistling teakettle in an adjoining exhibit. A colorful poster on the wall said, "Split atoms *and* wood." The purpose of the pleasant little museum was obviously to make people feel comfortable about nuclear power.

After completing a review of the exhibits, we donned hardhats, and entered the plant, stopping for several security checks, and passing through machines designed to detect explosives and metals. As I had discovered during several weeks of inquiries, it was not easy to gain admission to a nuclear power plant, so my interest in technical processes was augmented by the pleasant sense of entering a forbidden place. But when we stopped in an office labeled "Health Physics" and started filling out forms for the radiation safety department, my mood began to change. By the time I was dressed in a yellow plasticized paper coat, white cotton gloves, and yellow vinyl boots, carrying a film badge and a tubular dosimeter to record exposure to radiation, my cheerfulness had given way to apprehension.

We climbed steel stairs that clanged, passed through wire mesh doors opened with a computer key card, climbed more stairs, and came at last to the upper level of the fuel building. At one end of the large open room a number of metal cylinders were sunk into the floor, each one containing a fuel assembly which had just arrived at the plant. (The annual refueling was scheduled to take place within three weeks of my visit.) At the other end of the room was a pool of water containing spent fuel that had previously been removed from the reactor.

A fuel assembly is a column about 9 inches square and 12 feet tall, consisting of 204 slender stainless steel tubes each filled with pellets of uranium dioxide. At Connecticut Yankee, 157 such assemblies, each weighing about a ton, make up the total reactor "core." The core is located in the bottom of the reactor vessel, a steel chamber 41 feet high, 31 feet in diameter, with 10-inch-thick steel walls, which in turn is isolated in a huge concrete domed structure euphemistically called a "vapor container." Tony explained that we would not be going into the containment structure, since that was only entered once a week by special personnel to check on radiation levels. I did not complain.

During the annual shutdown, a third of the core is refueled, which means that 50 to 60 assemblies are replaced. (When one considers that a comparable coal-fired steam plant consumes more than a million tons of fuel each year, one reason for nuclear's allure becomes apparent.) The assemblies of spent fuel are moved from the reactor vessel to the storage pool through an underwater tunnel. Having served their purpose, they now exact a toll. They have become so fiercely radioactive that they must be carefully shielded. This is readily accomplished, at least for the short term, by keeping them immersed in the storage pool. Circulating water cools the material, and also serves as a shield against its radioactivity. Radioactive particles in the water are periodically removed by filtration.

After all that I had heard about the problems of accumulating nuclear waste, I was surprised to see how small the storage pool was: less than 37 feet square. Only its depth–35 feet–and the distortion this created, spoiled the illusion that it was a swimming pool in a neighbor's back yard. The pool at Connecticut Yankee has the capacity to store all the assemblies to be discharged by the reactor up to 1992.

As we stood looking down into the illuminated water, I asked my guide why none of the fuel was glowing as I had seen in photographs. He explained that all of this fuel was over a year old. "It loses more than 99 percent of its radioactivity during the first four months out of the reactor," he said. "Come back after the refueling, and you'll see this pool glowing like a television screen." Sensing my disappointment, he turned off the lights in the pool and most of the lights in the room, and started looking for assemblies that might still be glowing faintly. He finally found two near the center of the pool. A gantry crane spanned the water serving as a small footbridge, and we clambered onto that to get a better look. "Hold on to your hat," said Tony as I peered over the edge. This was not so easily done, since I found that I had a fierce grip on my pad and pencil and, particularly, the railing. I stared down into the depths and saw the pale blue glow of–what? the promise of a new Arcadia? the flaming eyes of Baal? In truth, I was not thinking in a poetic mode; I felt like getting out of the place.

I have visited factories and stood beside huge machines and roaring furnaces, usually with a feeling of exhilaration. Staring into the storage pool was a different sort of experience. I came away with a sense of the irrational dread–the *angst*–that underlies the widespread hostility to nuclear power. During the next few days, I told a number of people, including several engineers, that I had been inside a nuclear power plant. Almost without exception the news was received with a gesture or a remark that meant, "How exciting, but don't get too close to me."

When I left the radioactive part of the plant I was required to read my dosimeter and record on a chart the amount of radioactivity to which I had been exposed. With a mixture of relief and regret—like a war correspondent who has not heard a shot—I wrote a big zero next to my name.

It is an eccentricity of nature that atoms of U-235 (uranium with an atomic weight of 235) are unstable. When struck by neutrons, they split apart—fission—thereby liberating new neutrons, as well as large amounts of energy in the form of heat. Since uranium as found in the ground consists mostly of relatively stable U-238, and contains less than 1 percent U-235, the fuel for a nuclear plant is "enriched." This process consists of bringing the amount of U-235 up to between 3 and 4 percent, a proportion that has been found to be effective in sustaining a chain reaction. When the reactor core is filled with this enriched fuel, the fissioning U-235 releases heat. The heat is used to make steam that turns turbine-driven generators, as in an oil- or coal-fired plant. So far, so good.

Unfortunately, in the course of the chain reaction the U-235 breaks up into "fission products," lighter elements such as strontium-90, iodine-131, cesium-137, and cobalt-60. Also, some of the U-238 absorbs neutrons and is converted into heavier "transuranic" elements, especially plutonium-239. These sinister names have become familiar to us because they identify materials that are dangerously radioactive.

The radiation emitted by these substances is in the form of *alpha* (two protons and two neutrons), *beta* (high-speed electrons), and *gamma* (a high-energy form of electromagnetic radiation). When such particles or rays strike human tissue, they tend to split body cell molecules into electrically charged particles and chemically reactive fragments. These body cell parts, made alien by radiation, are unable to function normally. It is in this way that radioactivity can cause cancer or mutagenetic damage. Bombardment from sources outside the body or from radioactive particles that are ingested or inhaled can cause the damage. All of this is known to high school science students and readers of popular science magazines.

And if this were the whole story, any sane person would flee from radiation as from a plague, and nuclear power would long since have been banned. But debate arises because radiation is part of the natural environment, and certain amounts of it have always been assumed to be harmless. Since the body is continually replacing cells, it seems reasonable to assume that a certain number of cells can be damaged without ill effect.

As the concerned citizen has learned by now, the estimated biological effect of radiation on the human body is measured by the *rem* (for roentgen equivalent in man), or more usually by the millirem (a thousandth of a rem). According to statistics of the Environmental Protection Agency, the average American receives an annual dose of approximately 60 millirem from the natural radioactivity in soils, rocks, water, food, and air. (The naturally occurring elements that significantly affect humans are potassium-40, radium, thorium, and carbon-14.) In addition, cosmic radiation contributes about 40 millirem (the range is from 30 millirem at sea level to more than 100 in mile-high Denver). So just by living on the surface of the earth we are subjected to an average dose of 100 millirem. A single chest x-ray can add 100 millirem, a gastrointestinal tract x-ray 2,000. In all, the average annual dosage is about 200 millirem; for people who live at high altitudes and have more than one x-ray examination, it is obviously much greater.

Compared to this, a person who stayed at the border of a typical nuclear plant site 24 hours a day for an entire year would be exposed to no more than five millirem. A mile away, the figure becomes less than a single millirem, and more than five miles away the dose is usually undiscernible.

If the increase in radiation exposure to the general population is so negligible (and this holds true when we include other aspects of the nuclear cycle, such as mining, milling, and transporting), then why all the fuss?

Anxiety begins when sombody like Dr. John Gofmann, professor emeritus of medical physics at the University of California, announces that there is no safe level of radiation, and that the natural background radiation dose of 100 millirem actually kills some 20,000 Americans each year. This is the sort of remark that makes one want to start running, although there is obviously no place to hide. Gofmann then extrapolates his theoretical, unprovable assumption and announces that the nuclear industry "is causing an extra 2,000 cancer deaths a year."

This apocalyptic reasoning then becomes the basis for articles such as one that appeared in *Saturday Review* entitled, "Is Nuclear Power a License to Kill?" [18] It also accounts for an article in *The New Yorker* that tells of a woman who is afraid that her child might contract leukemia since they live "not far" from a route traveled by trucks carrying nuclear waste.[19] (The trucks emit less than one millirem per hour at a distance of three feet.) It is not said whether the woman permits her child to watch a couple of hours of television each day, for an annual dose of four millirem, or allows any dental x-rays at 20 millirem each.

It is only human to sympathize with the qualms of anxious mothers, but after reading arguments on both sides of the question, I concluded that the benefits of nuclear power more than offset the risk of exposure to an extra millirem or so of radiation, comparable to taking a short vacation in the mountains. As for the workers in nuclear plants, limited by government regulation to an annual exposure of five rem (5,000 millirem), all the evidence I could find indicates that their work is far less hazardous than that of most industrial workers, certainly less so than that of coal miners, 150 of whom die in accidents each year, to say nothing of the harmful coal dust to which they are exposed.

Of course, all of this reasoning rests on the assumption that there will not be a serious accident. There would be little comfort in government regulations that "protect" both workers and the general public should there ever be a mishap that releases more radiation than "permitted." This obvious concern led me to a review of the Rasmussen Reactor Safety Study, a three-year, $4 million study directed by Professor Norman C. Rasmussen of MIT and issued by the Nuclear Regulatory Commission in 1975. Using complicated reliability and safety-analysis techniques, the study concluded that the worst conceivable disaster (involving 3,300 fatalities, 1,500 latent cancers, 45,000 "early illnesses," and property damage of $14 billion) was likely to occur only once in a billion per reactor per year. The chance of an individual dying from a nuclear accident in any given year was estimated to be one in five billion. By comparison, the likelihood of being hit by lightning is on the order of one in two million.

Displaying these figures in literature as soothing as a resort brochure, and able to claim that no member of the public had ever been harmed by a nuclear plant, the Atomic Industrial Forum, in early 1979, was more than holding its own on the question of plant safety.

The nuclear conflict, however, instead of abating, had shifted to a new issue—the management of radioactive waste.

No group comes readily to mind that is less endearing than the utility companies, manufacturers, and suppliers which constitute the nuclear industry. But in their distress about radioactive waste, they do deserve a measure of sympathy, for the problem is not of their making, nor is the solution within their control. It was always anticipated that spent fuel from commercial light-water reactors, after cooling for a year or so in on-site storage pools, would be taken away to reprocessing plants, where new fuel would be reclaimed. Reprocessing consists of chopping up the spent fuel rods, dissolving them in nitric acid, and then subjecting the solution to chemical separation in which the uranium and plutonium are extracted for reuse. The residue liquid, a nasty brew that is acid and corrosive as well as radioactive, is called "high-level waste."

It is by this same process that materials for nuclear weapons are obtained. The main difference between military and commercial reactors is that in the former the heat produced by fission is a waste product; it is the plutonium that is wanted.

Unfortunately, the plutonium reclaimed in commercial reprocessing is also suitable for making atomic weapons, something that has troubled those officials charged with looking after American national security. In 1976 President Gerald Ford warned that non-proliferation objectives might lead to a ban on commercial reprocessing, and in April 1977 President Carter put such a ban into effect. Thus, the power industry was suddenly confronted with what has graphically, if inelegantly, been described as "the constipation problem."

For all of the industry's plaints about the need to "close the fuel cycle," however, reprocessing is not a solution to the "radwaste" problem. Although it takes the waste off the hands of the utilities (and conserves perhaps 20 percent to 40 percent of the uranium that is lost in the "once-through" method), it creates a waste liquid almost as radioactive as the spent fuel, and in some ways more difficult to manage. Spent fuel assemblies, at least, are packages that can be conveniently handled by crane. High level liquid waste has a dreadful history of leaking, and must be solidified in order to be effectively disposed of. Industry literature makes it sound as if a selected solidification technology is already in hand, but this is not so. The French government has built a vitrification plant that immobilizes the waste in glass, but not all scientists agree that this process is the most suitable.

The only privately owned fuel reprocessing facility ever to function in the United States was the notorious plant in West Valley, New York, operated from 1966 to 1972 by Nuclear Fuel Services, a subsidiary of Getty Oil. On many occasions it was accused of exceeding permissible limits on radioactive effluents and employees' exposure. Unable to operate effectively at a profit, the owners simply walked away, as they had a legal right to do, leaving the state and federal governments with 60,000 gallons of high-level wastes in leak-prone tanks. A second private reprocessing facility, constructed in 1974 by General Electric at Morris, Illinois, turned out to be a technical failure, and never went into production. A third facility was built in Barnwell, South Carolina, by Allied-General Nuclear Service, but the ban on reprocessing prevented it from opening.

Whether high-level waste is in the form of spent fuel assemblies or liquid waste vitrified into capsules, it remains lethal for hundreds of years, and so must be isolated from the biosphere. Several methods have been proposed, including disposal in deep-sea sediments or in outer space, but scientists seem to agree that the most sensible procedure is to deposit the stuff far underground in mined geologic repositories. The quantities involved are quite minimal—only about 33 tons of spent fuel per plant per year—and geological evidence indicates that in a properly selected medium, the procedure will be quite safe. Incredibly, however, more than three decades into the atomic age, the necessary research and development on this problem have not been performed.

In the beginning, high-level radioactive waste came exclusively from the making of weapons. Most of this material was stored at the site of military reactors in Hanford, Washington, and Savannah River, South Carolina. Smaller quantities accumulated at the Idaho National Engineering Laboratory, Idaho Falls, Idaho, where fuel from nuclear-powered submarines is reprocessed. The Atomic Energy Commission, from its birth in 1946, was responsible for managing these wastes. Its initial procedure was simply to neutralize the acid by adding sodium hydroxide (thereby doubling the volume) and then to store the liquid in large steel tanks. It was known from the start that some long-range solution would have to be found, but the matter was not addressed with any urgency.

Eventually, the AEC set about looking for a suitable underground site, but after two decades of bluster, blunder, and bad luck (scathingly chronicled in Philip Boffey's book *The Brain Bank of*

*America*), the agency in 1972 announced that as an interim measure it planned to build a storage facility on a government reservation at Hanford or the Nevada Test Site. This "mausoleum" installation would be suitable for use up to a hundred years.

But by 1972 the environmental movement had emerged, and the surface storage scheme was seen to be a quick fix that would allow further procrastination on a permanent solution. The fledgling Environmental Protection Agency expressed concern, and the AEC had to retreat.

As increasing quantities of spent fuel accumulated in nuclear plants, and as the military waste at Hanford was found periodically to be leaking from aged tanks, demands for action became more pressing. In 1976, AEC's successor, the Energy Research and Development Administration (succeeded in 1977 by the Department of Energy), embarked on an intensive search for underground repository sites, and asked the cooperation of 36 states in this enterprise. The response was, to say the least, hostile. Michigan, New York, and Ohio—three states that generate a large percentage of the nation's nuclear waste and also happen to sit on top of major deposits of salt (considered a likely repository material)—all refused to allow studies to proceed. Eight states promptly legislated bans against repositories, and another dozen made plans to do the same. The new populism of the 1970s required that the Federal government correct by consensus what the once-haughty AEC had wrought by fiat.

Near the end of his first year in office, President Carter created an Interagency Review Group "to formulate recommendations for establishment of an administration policy with respect to long-term management of nuclear wastes." The group's draft report, issued in late 1978, recommended a program of research, development, and construction that would achieve a functioning repository within 10 to 17 years, depending upon how much emphasis was put on preliminary research.

The IRG report was a masterpiece of bureaucratic prose, described in *Science* magazine as "technically conservative and politically discreet." It was received without animosity—indeed, with respect—by both the nuclear industry and its adversaries. As for the politicians, Governor James B. Edwards of South Carolina, chairman of the Nuclear Power Subcommittee of the National Governors' Association, commended the IRG for taking a "most unbureaucratic and refreshingly productive" approach.

The report succeeded by being disarming. Although far from being a *mea culpa,* it admitted past errors and present uncertainties, and it emphasized a concern for public opinion and regard for the rights of local government. It proclaimed neutrality on the question of reprocessing and indeed, on the future of nuclear power itself, maintaining that whether the problem is spent fuel or postreprocessing waste, whether a rapidly accumulating amount or merely the material on hand, plans must be made for disposing of the stuff satisfactorily. Nobody could quarrel with such an approach.

But when I read the report I felt as if I were trying to evaluate wall posters in Peking. Behind the subtle allusions lay harsh political realities, a power struggle between several federal agencies, conflict between local and federal authority (which the report shied away from with an appeal to something called "cooperative Federalism"), arguments between scientists about how fast to move and in what direction, and the violently disputed question of whether the licensing of new nuclear plants should be "linked" to progress in waste management.

The initial printing of the IRG report, 7,000 copies, was not nearly enough to satisfy demand, and eventually 15,000 were distributed. More than 3,300 written comments were submitted by groups and individuals. As applause for the report died down, the contesting parties intensified their lobbying efforts, and a feeling of inertia gradually returned.

With the basic facts in hand, I set about writing my article for *Harper's.* Since questions about plant safety appeared practically resolved, I concentrated on the issue of nuclear waste, and reported what my studies had revealed, that this was less a technical problem than a matter of political paralysis, and that what was needed was firm direction from Washington.

Recognizing my own limitations as a non-expert in the field, I relied not only on my own evaluation of the available literature, but also on interviews with nuclear physicists and geologists—caring, sensitive, family-loving individuals with no personal stake in the nuclear debate. I was confident that I had good information and that my presentation was responsible.

The first draft of my article was heavy with agency initials—AEC, ERDA, DOE, NRC, EPA, IRG, OSTP, OMB, and so on—and further burdened by a none-too-clear survey of peripheral disposal problems such as low-level nuclear waste, uranium mill tailings, transportation, and decommissioning of old plants. The essay was returned to me for some much-needed reorganizing and clarification.

At about this time—late January 1979—the Nuclear Regulatory Commission unexpectedly endorsed a critical analysis of the Rasmussen Report, and the long-dormant question of nuclear plant safety became once again a live issue. Professor Rasmussen acknowledged that his calculations might have been too optimistic by about a factor of ten, and although this did not make a significant difference in the report's conclusions, public perception was quick to change. I found myself rewriting my article to take account of increasingly alarmist headlines.

I could not, however, keep pace with events. The biological effects of low-level radiation—a subject about which little had been heard for some time—suddenly appeared on the program of several scientific meetings. Although there were no significant new facts, there were angry debates that received wide coverage in the press. Then the movie *The China Syndrome* opened, and a nervous tremor seemed to sweep across the nation. On March 13, the NRC closed down five nuclear plants because of errors discovered in the design of certain pipelines, and just as I was trying to figure out the significance of this, on Wednesday, March 28, the crisis began at Three Mile Island.

Early the following Saturday morning I received a telephone call from Lewis Lapham, the editor of *Harper's*. Had I been watching the news programs on television? he asked. Had I seen what was happening? Could I rework my article to take account of this mass hysteria?

Rework it I did, all the rest of the weekend. But I was struck by a crippling ailment—the lack of a point of view. The technical facts had not changed. The Rasmussen Report was not discredited by the events at Three Mile Island. If anything, its assumptions were proved valid. Nuclear power was still "safe"—at least statistically safer than any other source of electrical power available for the near future, and also environmentally more acceptable.

While everyone was cringing over events at Three Mile Island, airborne effluents from coal-fired power plants continued to drift across the land carrying poisonous mercury, cadmium, lead, and sulfates—and (because American coal contains uranium and thorium), substantial doses of radiation. Millions of tons of ash and toxic sludge continued to accumulate. Nothing had happened to change the conclusions to be found in the 1977 Ford Foundation-sponsored report, *Nuclear Power Issues and Choices:* even new coal-fueled power plants meeting new environmental standards "will probably exact a considerably higher cost in life and health than new nuclear plants." [20]

The scientific facts had not changed, but I found myself unable to discount the feelings of people all around me. Widespread dread—this also was a fact. Clearly a society has the right to choose a greater number of probable deaths (through increased use of coal) in exchange for a reduction of anxiety (through decreased use of fission). Conceivably people might choose a more primitive way of life, even a higher death rate—fewer ambulances, hospitals, and medical equipment, a lower gross national product—to eliminate the dread of a nuclear accident, no matter how unlikely.

These are not choices that I approve–they do not *make sense*–yet during the Three Mile Island crisis I could not bring myself to disparage them.

The final version of my article was a litany of qualifications. "It's weak," Lapham said ruefully. I could not disagree, and the piece was shelved. It is difficult to write for a journal of opinion when one is suddenly bereft of robust opinion. Not that *Harper's* wanted me to take one particular side or the other. Their mission is to provide a forum for stimulating debate, and I believe that they would have welcomed either a vigorous, well-reasoned defense of nuclear energy in the face of public panic or a scathing denunciation of the nuclear industry's performance as exposed by the near-disaster. They might also have been satisfied with a judiciously balanced piece, but it would have had to be extraordinarily good, since judiciousness does not sell magazines.

Nietzsche's Zarathustra reserves a special scorn for those who take a position in the *midst*. "That," he declaims, "is *mediocrity, though it be called moderation.*"

"Nuclear Disaster," read the cover of *Time* magazine the week after Three Mile Island. Disaster indeed it was for me, with several months work gone up in smoke. And disaster it was for the nuclear industry, although probably not a fatal one. Nevertheless, once the danger had passed, I found it possible to view the crisis with a degree of somber satisfaction.

The events of 1979–a year that started with Three Mile Island and ended with American hostages being held by inscrutable fanatics in Iran–demonstrated that the future of nuclear power depends at least as much upon human needs and human fears as upon technical considerations. Government investigations, lawsuits, public demonstrations, a plethora of articles, editorials, and letters to the editor–all contributed to a frenzied atmosphere in which appeals to probabilities and statistics went virtually unheard. Engineering problems were transmuted into questions of bias and predilection, even doctrine and heresy. However disquieting this may be, the very irrationality that has come to dominate the nuclear debate confirms that the public will is still what counts.

The apprehensions of ordinary people are *factored into* the decisions that are made. They show up in the form of stricter regulations by government agencies, delayed approvals by cautious politicians, cancelled orders by uncertain utility executives. These in turn result in an increased cost for nuclear power, and consequently decreased use. Fear of the atom also is reflected in larger government grants for development of alternative energy sources, and growing interest on the part of researchers and investors in these alternative sources. On the other hand, the rationally, statistically demonstrable advantages of nuclear power are also part of the ever-bubbling brew out of which comes action in a republic.

Nobody can predict what is going to happen first: a serious accident in a nuclear plant, a revolution in Saudi Arabia or, on a happier note, a breakthrough in atomic fusion or solar cell technology. But I am satisfied that many citizens together–blending sense and instinct, boldness and caution–are making the choices that must be made. The antitechnologists mourn the passing of a democracy that is very much alive. The human spirit still calls the tune.

# 7

# Muddled Heads and Simple Minds

Anyone who has attempted to defend technology against the reproaches of an avowed humanist soon discovers that beneath all the layers of reasoning–political, environmental, aesthetic, or moral–lies a deep-seated disdain for"the scientific view." Because scientists are taught to eliminate personal bias from their frame of reference, the humanist assumes that such people have no access to important "truths," and that they see, to use Theodore Roszak's graphic phrase, "with a deadman's eyes." If technologists and humanists are not simply to "talk past each other," this charge must be confronted.

In *The Existential Pleasures of Engineering* I endeavored to defend the practice of engineering as a fundamentally human pursuit, one that springs from the wondrous sources of craftsmanship. But here the task is different and more challenging: to argue that science (the scientific method, the scientific view) need be no less human (sentient, tender, aware) than any other mode of perception (literary, poetic, artistic). This may seem like an unnecessarily deep place from which to start a discussion of technology, but unless this point can be established, technologists will continue to be patronized whenever the discussion gets down to "basics," and our most carefully reasoned discourse is likely to be rebuffed out of hand.

Joseph Wood Krutch was a professor in the English Department of Columbia University when I was a graduate student there during the academic year 1946–47. One of Mr. Krutch's specialties was "the classical period" in English literature (roughly the latter half of the eighteenth century); in particular, he was noted for his lectures on Samuel Johnson, whose biography he had written in 1944. I enjoyed attending those lectures, and I found the study of Johnson and his contemporaries to be a refreshing antidote to a program that was otherwise heavily laced with works of the Romantic poets. In addition to teaching at Columbia, Krutch was drama critic for the *Nation.* I have never encountered, in any classroom, a more urbane, acerbic raconteur. In many of my other courses, I was being taught that great literature comes from the heart; but from Joseph Wood Krutch—and Samuel Johnson—I acquired an enduring respect for wit, intellect, and common sense. I also greatly admired Krutch's book *The Modern Temper,* which had become something of a mini-classic since its publication in 1929. In it Krutch considered the ways in which science had stripped man of his credentials as God's chosen representative in the cosmos, and while the book was noted for its "pessimism," I was inspired by its tone of courageous stoicism.

Krutch retired in 1950, at the age of 57, partly for reasons of health, and moved to Arizona, where he lived in an adobe house in the desert and embarked on a new career as an amateur naturalist. He wrote prolifically, and gained a large and appreciative audience. I followed his work closely and read each of his books in turn: *The Desert Year* (1952), *The Best of Two Worlds* (1953), *The Measure of Man* (1954)—which won a National Book Award—*The Voice of the Desert* (1955), *The Great Chain of Life* (1957), *Grand Canyon* (1958), and *Human Nature and the Human Condition* (1959). Krutch wrote several other books before his death in 1970, but those just named were his main contributions to the philosopher-naturalist literature.

They were all wonderful books, well-crafted and genial, but I was astonished to see that as the years went by, this man whom I had so admired for his unflinching lucidity was becoming increasingly hostile to the scientific view. The 1950s was the decade of the Cold War, McCarthyism, the Korean War, and the hydrogen bomb, a time in which the phrase "age of anxiety" first gained currency. I do not know how much of Krutch's changing outlook was attributable to the events of the day, how much to his studies of life in the desert, how much to his own advancing age or to other influences. But I do remember how distressed I was to see this shrewd, witty son of the classical age slip into a murky, anti-scientific mood. In his sensitivity to the importance of ecological balance, he was an admirable pioneer, almost a generation before his time. But his philosophical pilgrimage away from the Enlightenment was also an augury of things to come, a drawing back from truth that I reluctantly came to recognize as a failure of nerve.

In *Human Nature and the Human Condition* Krutch outlined his disillusionment with science in the following terms:

Today the prevailing opinion among even the moderately intelligent and instructed is based largely upon their understanding and misunderstanding of Darwin, of Marx, of Freud, and, more especially, of their popular expositors. From the teaching of these masters they conclude: (1) that man is an animal; (2) that animals originated mechanically as the result of a mechanical or chemical accident; (3) that "the struggle for existence" and "natural selection" have made man the kind of animal he is; (4) that once he became man, his evolving social institutions gave him his wants, convictions, and standards of value; and (5) that his consciousness is not the self-awareness of a unified, autonomous *persona* but only a secondary phenomenon which half reveals and half conceals a psychic nature partly determined by society, partly by the experiences and traumas to which his organism has been exposed.

Thus though man has never before been so complacent about what he *has,* or so confident of his ability to *do* whatever he sets his mind upon, it is at the same time true that he never before accepted so low an estimate of what he *is.*[21]

The concept of man as an animal could not have been, in itself, troubling to Krutch. As an eloquent nature writer, he had convincingly shown how admirable are many animal traits and how comforting and significant is the bond between natural man and his fellow creatures.

At the root of Krutch's unhappiness, apparently, was the positivist view that human beings and animals are composed essentially of chemical "matter," a view that Krutch held to be not only harmful and dismaying in its effects, but absurd in its assumption. How could a human, that fabulous, value-making, beauty-seeking, living creature of conscious awareness be composed solely of "matter"?

But here Krutch reminds one of the policeman accosting a burglar who had just broken into his own house by mistake. The scientist has been caught with the goods but must be declared innocent. He has merely broken into his old uncertainty through the rear door of his new truth.

The basic "stuff" of the universe, as it is represented by today's science, is so paradoxically inscrutable in both form and behavior that we can, without doing violence to the facts, regard it as just about anything people have ever thought it to be–life force, will, mind, idea, dream, spirit, God (or the mind of God), or, in the equally obscure terms of science, electricity, gravity, field, or a "warpage" of space (whatever *that* is). However we conceive of it, *it is what it is,* essentially vital and virtually limitless in potential.

Is it presumptuous to suggest that the humanist's problem may be principally a question of numbers? After all, the human brain, which is itself a product of evolution, is not equipped to think in terms of large numbers. We can readily visualize three apples, four, five, six, seven . . . but as the number increases, we falter and a feeling of inadequacy, even dizziness, overcomes us. How are we, then, to imagine the possible creations of an infinite number of atomic particles active over a period of several billion years? How are we to conceive of a human being comprised of more than a trillion living cells, each cell with its tiny nucleus of 46 chromosomes, each chromosome containing thousands of genes, each gene consisting of about a million atoms?

How are we to visualize the potentialities of this organism in action? If we spend five minutes watching a baby in its crib, we see an almost infinite number of acts of learning, observing, and experimenting. It seems like an eternity to us, and must certainly seem like one to the baby also. When we consider that the baby experiences tens of thousands of these eternities in its first year, we begin to get the faintest notion of the mathematics of living.

Inconceivable myriads of events in space–time are around us and within us. Even if each event were simply identifiable in scientific terms, the whole would overwhelm us. Like Pascal, as we probe into the abysses of the infinitely great and the infinitely small, we "tremble at the sight of these marvels." There are marvels and potentialities aplenty in the scientific world view without romantic philosophical embellishments.

However, even if we were able to bludgeon Joseph Krutch with every scientific fact known, and with the many that will soon be discovered, he might retreat only to take his stand on new ground. Logical truth aside, this Jamesian pragmatist might say, it is harmful for man to believe in positivism, evolutionism, and relativism; therefore these are not *good* truths. But as Mr. Krutch so clearly knew, or appeared to know before his retirement to the desert, we have come too far to start confusing the true with the pleasant. We cannot regain our intellectual virginity or, to use one of humanity's most poignant metaphors, pretend that we have not tasted the fruit of knowledge.

What is even more to the point, we need not admit that as humans discover that their values are relative, they must also assume that the values are worthless. We may *choose to affirm* whatever we find congenial to our nature, recognizing that we were human beings long before we became scientists.

A person may study the biological origin of love, yet love none the less ardently. We may understand how religion grew out of the fear and ignorance of primitive peoples and flourished because of the survival value of its moral precepts, yet remain receptive to religious experience. We may accept the fact that the cosmos is indifferent to our values and still remain loyal to them; be aware that all is relative and yet at certain moments cry out that evil is absolute, that beauty is truth. We may know that our intuitions are suspect, yet prize them nevertheless, acknowledge that ultimately our will is not free, yet act as if it were. We have the need and, happily, the ability to juggle two world views, one objective, the other subjective, and we disregard either at our peril.

A most striking example of these two states of mind is found in the sexual urge. When this impulse is ascendant, the objective intellect becomes a dim voice indeed, carrying little persuasive power. But a few instants after gratification a radical and mysterious change occurs, making cool detachment possible once again.

Such alternate states of tension and release may hold the key to many a dispute between the humanistic and scientific philosophies. The "unclouded and attentive mind" that Descartes relied on so naively we now know to be only rarely attainable, comparable in a way with the states of ecstasy that lie at the other end of the human spectrum.

Instead of accepting and valuing both, or all, states of being, Krutch made his choice exclusively for sublime tension. He believed "the accounts of the poets, novelists, and playwrights to be truer" than those of the scientists. He objected to our increasing "knowledge about" things and wished we had more "knowledge of," a direct awareness of "things in themselves."

About the quest for such sensual "knowledge," Santayana has made this wry comment: "Knowledge is not eating, and we cannot expect to devour and possess *what we mean*. Knowledge is recognition of something absent; it is a salutation, not an embrace." [22]

And truly we have little reason to believe that the intuitive "embrace" of life is the answer to modern man's dilemma. What would Krutch have gained if he could have saved people from becoming robots only to see them turn into hippies? Human beings have worshipped often at the altars of art, mysticism, and sensuality without finding salvation. What new developments have occurred to induce them to turn their backs again on the hard world of "things" and "doing"?

Goethe's Faust, as I have noted, did not find his moment of ultimate fulfillment until he became a civil engineer building dikes. Kafka's K. hoped to achieve a state of grace by obtaining work as a land surveyor. Kafka, incidentally, was fond of repeating an anecdote concerning Gustave Flaubert, who in his later years happened to spend a day in the country with a bourgeois family wholly without intellectual or artistic pretensions. On his way

home this great writer, who had sacrificed his life to his art, kept
thinking of these simple people with their simple pleasures, and he
muttered over and over, *"Ils sont dans le vrai!"*

To be *dans le vrai*—in the right—is what we are all after, of
course, scientists as well as humanists, and if we have learned anything at all in our generations of searching, it is that there is no
single road to the good life, or to the bad life, for that matter.
Certainly the cycle of human oscillation between opposite philosophical poles precedes by centuries the ascendance of science. And
it has not been demonstrated, in spite of recent humanistic cries of
alarm, that the scientific view of the universe is dehumanizing
humanity, if anything is.

We can sympathize with Krutch's vexation and uneasiness
over a handful of self-important scientists and would-be social manipulators. But in trying to attack these menaces at the roots, so to
speak, he put himself in the position of defending metaphysical
propositions that are becoming daily more untenable.

Every thinking person who believes in the truths of science
undergoes, sooner or later, emotional experiences that call these
truths into question. Bertrand Russell, another intellectual hero
of my youth, tells of undergoing a spiritual crisis during World
War I:

> I used to watch young men embarking on troop trains to be
> slaughtered on the Somme because generals were stupid. I
> felt an aching compassion for these young men, and found
> myself united to the actual world in a strange marriage of
> pain. All the high-flown thoughts that I had had about the
> abstract world of ideas seemed to me thin and rather trivial in
> view of the vast suffering that surrounded me. The non-hu-
> man world remained as an occasional refuge, but not as a
> country in which to build one's permanent habitation.

Russell, however, unlike Krutch, refused to draw back from his commitment to the scientific view:

> My abandonment of former beliefs was, however, never complete. Some things remained with me, and still remain; I still think that truth depends upon a relation to fact, and that facts in general are non-human; I still think that man is cosmically unimportant, and that a Being, if there were one, who could view the universe impartially, without the bias of *here* and *now,* would hardly mention man, except perhaps in a footnote near the end of the volume.[23]

Russell's lifelong involvement in political causes, as well as the turbulence of his own romantic life, are ample proof that an acceptance of the truths of science in no way lessens one's quotient of human passion.

The story is told of a chance encounter between Bertrand Russell and Alfred North Whitehead. The two had worked together as young mathematician–philosophers in the writing of the massive *Principia Mathematica,* after which each had gone his own way. Russell had maintained his faith in science, but Whitehead's work gradually took on transcendental overtones. "Alfred," Russell is supposed to have said, "I have been reading some of your recent works, and I'm sorry to see that you have become terribly muddle-headed."

"Perhaps," came Whitehead's reply, "but you, Bertie, seem to have remained rather simple-minded."

We all tend to become either muddle-headed or simple-minded—or, to use William James's classifications, either tender-minded or tough-minded. But this does not mean that we cannot understand each other, talk to each other, and even occasionally convince each other that one course of action is to be preferred to another.

Mustering such confidence as we can for the possibilities of communication between people of differing intellectual dispositions, let us continue.

# 8

# Small Is Dubious

hile reading the newspapers the morning after President Carter's energy address to the nation in April 1979, I was struck by a statement attributed to Mr. Carter's pollster and adviser, Patrick Caddell: "The idea that big is bad and that there is something good to smallness is something that the country has come to accept much more today than it did ten years ago. This has been one of the biggest changes in America over the past decade."

Since the nation had just been exhorted to embark on the most herculean technological, economic, and political enterprises, the reference to smallness seemed to me to be singularly inapt. Waste is to be deplored, of course, as is inefficiency. But bigness? I had not realized that the small-is-beautiful philosophy had reached the White House.

Of all the ideas that have emanated from the opponents of technology, none has been more confusing and potentially dangerous than "small is beautiful." The early literature of anti-technology, from Jacques Ellul's *The Technological Society* (1964) through Theodore Roszak's *Where the Wasteland Ends* (1972), for all its faults, was relatively harmless. It advocated a passive resistance to high technology, or at most a "withdrawal" by sensitive souls from the prevailing culture. The publication in 1973 of E. F. Schumacher's book *Small Is Beautiful* heralded a more aggressive mood. Schumacher and his disciples are not content to brood. Their doctrine of smallness advocates a headlong retreat into the past, or worse, a heedless rush into a future that will not work. This new movement is all the more to be feared because it has such a seductive rallying cry.

The initial success of Schumacher's book was in some respects deserved. It could not help but strike a responsive chord in anyone who was familiar with many abortive attempts to transfer high technology to the developing nations. Undoubtedly some foolish, insensitive things were done, and these deserved to be criticized. (On the other hand, some of the "small" technologies introduced into nations such as India have also been fiascos—for example, wind-powered water pumps installed in areas where there are long seasons of windlessness, bio-gas generators that fail to work in the cool of winter, and specially subsidized village soap industries that collapse in competition with factory-produced brands.)

It is not my purpose, however, to argue on behalf of any

particular master plan for India or any other Third-World nation. Nor do I suggest that every society model itself after the United States. If Schumacher had restricted himself to the topic of technology transfer from developed to undeveloped lands, his contribution would have been interesting and responsible, even if not altogether convincing. Unfortunately, with convoluted missionary zeal he transmuted his approval of primitive village life into a critique of Western industrial society.

When Schumacher's book arrived on the scene in 1973, the United States was reeling under the effects of an oil embargo, an environmental crisis, the just-ended Vietnam War, and Watergate. Our problems were formidable, and our large institutions seemed to be showing signs of strain. Somehow the phrase "small-is-beautiful" had a soothing ring. It sounded like an idea whose time had come. In fact, it is an idea that does not bear scrutiny, and if it takes hold in our thinking, it has the potential for doing much harm.

To an engineer in the United States, the debate about whether technologies should in principal be large or small, hard or soft, high or low, is almost incomprehensible. Even when the word *appropriate* is used, the arguments seem absurd. Of course, everyone agrees that technologies should be "appropriate"–but to define "appropriate" as "small" or even "intermediate," as so many people are doing, is to beg the question. When engineers are confronted with a problem of design, in which the objective is to satisfy certain human needs or desires, often the solution that appears to be most effective, economical, elegant, satisfactory, suitable, fit, proper, *appropriate*–call it what you will–is a technology that is large in scale.

Take, for example, the use of water. One of the oldest technologies–indeed a technology that is associated with the beginnings of civilization–is the control and distribution of water, the *sharing* of water by large numbers of people. Obviously our society could not survive without large-scale study and allocation of our water resources, without reservoirs, dams, water treatment plants, pumping stations, canals, and aqueducts. It is not possible for each small group of people to drill a well or divert a stream. Engineers have not arbitrarily decided to make water distribution a large technology; their engineering solutions have been inherent in the very scheme of things. Other technologies are also by their nature "large"—railroads, highways, airports—most of the undertakings that fall into the category of what we call public works.

This does not mean that engineers are committed only to large technologies, nor that they are mesmerized by massive projects. A tiny hearing aid or a pacemaker or a small transistor radio is just as exciting an engineering achievement as a bridge or a dam or a Boeing 747. Engineers are constantly trying to evaluate the appropriate scale for their designs. Most households have their own small refrigerators making their own small quantity of ice. It would be no great trick to make the ice at a central location and distribute it to each home by vacuum tube, or even to fly ice in from the arctic and drop it at each doorstep by helicopter. However, we seek the system that seems to "make sense," including economic sense.

In the design of tools and machines, as in the design of systems, engineers are also pragmatic. A gargantuan printing press seems right for a weekly magazine with a circulation of several million people; a small press or copying machine is appropriate for a local stationer; and for artisans who do etchings or woodcuts, there are beautifully designed and crafted hand tools. A thousand similar examples could be given, but there is no need to belabor the obvious.

Often we find that large technologies and small are intertwined. Consider the backpack—small technology, the very essence of the counterculture life-style—yet this item is made of aluminum and nylon, products of two very large energy-intensive, mass-production technologies. Any small technology that uses metal—say, the bicycle—is dependent upon the large technologies of mining and metallurgy.

A concern for "appropriate" technology would seem to call for both large and small solutions in an ever-changing creative flux. The world around us demonstrates a need for both large and small—whales and plankton, oceans and pools, redwood trees and wildflowers, huge migrating flocks and tiny isolated organisms. By demonstrating flexibility and variety, we fit naturally into the biosphere. Why, then, has an impassioned debate arisen over the question of small technologies versus large? Why do we find hostility directed against so many of the things that engineers build and deem worthy: dams, highways, power networks, large systems of all sorts?

Certainly the small-is-beautiful concept cannot be argued from a purely technical point of view. Our large systems have been, and are, technically successful. They are not perfect; they are merely superior to other systems presently available.

It is Schumacher's contention that a nation of small communities, populated by self-reliant craftsmen, is not only socially desirable, but also economically efficient. He said this in his writings, and also in a series of lectures, one of which I heard him deliver shortly before his untimely death in 1977. This idea sounds fine in a symposium on the human condition, and it was enthusiastically received by his audience; but the concept overlooks the enormous practical and psychological difficulties that stand in its way. Anyone who walks through a large city can see that the life-support systems for millions of people, many of whom are socially and educationally disadvantaged, cannot derive from neighborhood gardens, community bakeries, and roof-top solar panels. Even the modest government initiative in urban homesteading, in which abandoned homes are turned over to families at no cost, foundered because only a select few people have the technical knowledge and business acumen needed to restore an abandoned house to livable condition.

We could, of course, try to reduce our dependence upon high technology by doing away with large cities altogether, and move everybody "back to the earth." Such an experiment in self-reliance has been tried in our time by the Cambodian Communists, and what ensued was one of the great calamities of human history.

Neither Schumacher nor his followers have seriously advocated such apocalyptic designs. They would be willing to let us keep our large cities if only we could dispense with the enormous technological and social systems upon which these cities depend.

A prime target of their criticism has been the nationwide grid of electrical power. Amory B. Lovins, a British physicist, became an instant celebrity when he dealt with this subject in an article entitled "Energy Strategy: The Road Not Taken?" which appeared in the October 1976 issue of *Foreign Affairs.* The article, which argued forcefully for the development of small, or "soft," energy technologies, was extensively quoted in the international press, entered into the Congressional Record, and discussed in *Business Week. Foreign Affairs* received more reprint requests for this article than for any other they had published.

The existing "hard" system for distributing electrical power has deficiencies that are obvious enough. Electricity is created in huge central plants by boiling water to run generators. Whether the heat that boils the water is furnished by oil, coal, gas, atomic power, or even by a solar device, a great deal of energy is wasted in the process; an additional amount is lost in transmission over long lines. By the time the electricity arrives in our home or factory and is put to use, about two-thirds of the original energy has been dissipated. Also, the existence of the infrastructure of the power industry itself—tens of thousands of workers occupying enormous office complexes—costs the system more energy, and costs the consumer more money.

The alternative proposed by Lovins is the creation of small, efficient energy-creating installations in the buildings where the energy is used or, at most, at the medium scale of urban neighborhoods and rural villages. Direct solar plants are the preferred system, although Lovins also mentions small mass-produced diesel generators, wind-driven generators, and several other technologies still in the development stage.

Whatever the advantages of this concept (and Lovins's technical calculations have been challenged by a host of experts), its widescale adoption would obviously entail the manufacture, transport, and installation of millions of new mechanisms. This cannot but be a monumental industrial undertaking requiring enormous outlays of capital and energy. Then these mechanisms would have to be maintained. We all resent the electric and phone companies but, when service is interrupted, a competent crew arrives on the scene to set things right. It is easy to say that the solar collectors or windmills in our homes will be serviced by our independent neighborhood mechanic, but this is a prospect that must chill the blood of anyone who has ever had to have a car repaired or tried to get a plumber in an emergency.

The critics of high technology are fond of pointing out that large systems are "vulnerable." When there is a blackout, it is apt to be widespread. When there is an earthquake or a hurricane—or a war or sabotage—the potential for extended damage is great. This is true; but the vulnerability of large systems to large failures is more than offset, I believe, not only by the benefits of civilization being worth the risks, but by the self-healing capability of large communities. Cities that have been half-leveled by earthquakes or floods are miraculously regenerated in astonishingly brief periods of time, except that *miraculously* is the wrong word to use. It is large organizations, not miracles, that make it possible to tend to the injured, care for the homeless, and rebuild the metropolis. In primitive communities sufferers from storm, earthquake, or avalanche are left to scrape at rubble with little more than their bare hands, unless, of course, help is sent from the big cities. Also when struck by plague, drought, or famine, the residents of poor, rural areas simply suffer and die—unless, again, rescue missions are mounted by the much-maligned centralized bureaucracies.

According to Schumacher, Lovins, et al., large bureaucracies are uneconomical. If, for example, we could do away with the administrations of the large utilities companies, we would realize great savings. This is the age-old dream of "eliminating the middleman." But in the real world it appears that the middleman does perform a useful function. How else can we explain the failure of the cooperative buying movement, which is based on the idea that people can band together to eliminate distribution costs? The shortcomings of large organizations are universally recognized, and *bureaucratic* has long been a synonym for *inefficient.* But, like it or not, large organizations with apparently superfluous administrative layers seem to work better than small ones. Chain stores are still in business, while mom and pop stores continue to fail. Local power companies, especially, are a vanishing breed. Decisions made in the marketplace do not tell us everything, but they do tell us a lot more than the fantasies of futurists.

This is not to say that the situation cannot change. If a handy gadget becomes available that will heat a house economically using wind, water, sunlight, or moonlight, most of us will rush out to buy it. On the other hand, if the technological breakthroughs come in the power-plant field—perhaps nuclear fusion or large-scale conversion of sunlight to electricity—then I, for one, will be pleased to continue my contractual arrangements with the electric company.

Such an open-minded approach has no appeal to Lovins. Quoting Robert Frost on two roads diverging in a wood, he asserts that we must select one way or the other, because we cannot travel both. The analogy is absurd, since we are a pluralistic society of almost a quarter billion people, not a solitary poet, and it has been our habit to take every road in sight. Will it be wasteful to build power plants that may soon be obsolete? I think not. If a plant is used for an interim period while other technologies are developed, it will have served its purpose. When billions of dollars are spent each year on constantly obsolescing weapons that we hope we will never have to use, it does not seem extravagant to ask for some contingency planning for our vital energy systems.

Our resources are limited, of course, and we want to allocate them sensibly. At this time it is not clear whether the most promising energy technologies are "hard" or "soft" or, as is most likely, some combination of both. The "soft" technologies are not being ignored. Research and development funds are being granted to a multitude of experimental projects. At the same time, we are working on improvements to our conventional systems. What else could a responsible society do? We must assume that the technologies that prevail will be those that prove to be most cost-effective and least hazardous. Lovins claims that political pressures are responsible for government support for hard technologies, particularly nuclear. There may be some truth to this, but such pressures have a way of cancelling each other out. A new product attracts sophisticated investors, and before long there is a new lobbyist's office in Washington. The struggle for markets and profits creates a jungle in which the fittest technologies are likely to survive.

Technological efficiency, however, is not a standard by which the small-is-beautiful advocates are willing to abide. Lovins makes this clear when he states that even if nuclear power were clean, safe, and economic, "it would still be unattractive because of the political implications of the kind of energy economy it would lock us into." As for making electricity from huge solar collectors in the desert or from exploiting temperature differences in the oceans or from solar energy collected by satellites in outer space–these also will not do, according to Lovins, "for they are ingenious high-technology ways to supply energy in a form and at a scale inappropriate to end-use needs." Finally, he admits straight out that the most important questions of energy strategy "are not mainly technical or economic but rather social and ethical." [24]

So the technological issue is found to be a diversion, not at all the heart of the matter. We can show a thousand ways in which large technologies are efficient and economical, and it will have availed us nothing. The *political* consequences of bigness, it would appear, are what we have to fear. A centralized energy system, Lovins tells us, is "less compatible with social diversity and personal freedom of choice" than the small, more pluralistic approach he favors.

But is it not paranoiac to speak of losing our political freedom by purchasing electrical power from federally supervised utility companies? Is there any evidence to show that large technological systems lead to political despotism? More than 200 years ago Americans agreed to give up their individual militias and entrust the national defense to a national army. Was that a mistake? Has it led to a police state? Of course not. Diversity and freedom in the United States are protected and encouraged by strong central institutions.

I grew up on a diet of Western movies, and I cannot forget the small communities in which nice families were constantly being harassed by bad guys. Help invariably arrived in the form of the U. S. marshal, who came from the "big" society to protect people in the "small" society from the predations of their neighbors. In my adult life I have found that the situation has not changed much. Exploitation thrives in small towns and in small businesses. The personal freedoms that we hold so dear are achieved through big government, big business, big labor unions, big political parties, and big volunteer organizations. We want as much independence as we can get, of course; but independence flourishes best within the protective framework of a great national democracy. This is the beautiful paradox of America.

When large organizations challenge our well-being, as indeed they do—monopolistic corporations, corrupt labor unions, and so forth—our protection comes not from petty insurrections, but from that biggest of all organizations, the federal government. And when big government itself is at fault, the remedy can only be shake-ups and more sensible procedures, not elimination of the bureaucracy that is a crucial element of our democracy. Surely we would not be freer citizens if we made our own electricity or pumped our own water, much less if we had to buy either from our friendly neighborhood utility. Small-is-beautiful makes no more sense politically than it does technologically.

All right, say the small-is-beautiful advocates, even if large technologies "work" technically and politically, they do not work socially. The subtitle of *Small Is Beautiful* is "Economics As If People Mattered." Only in small social groups, according to Schumacher, is it possible for people to lead worthwhile lives.

Is this so? Not if there is any truth in such books as *Winesburg, Ohio; Spoon River Anthology;* and *Main Street,* with their picture of the American small town as a petty, cramped, and spiteful community—nor if you look through *Wisconsin Death Trip,* a documentary study of life in the Wisconsin countryside between 1880 and 1910—a journal of disease, early death, drudgery, insanity, suicide, arson, hostility, and despair that is nothing less than horrifying to read.

Big cities have their shortcomings, to be sure, but they have nurtured so much of what is estimable in human culture that to preach small over large at the social level reflects a shocking lack of warmth and sensibility. Moreover, big cities enhance small-town life, not only by providing a source of cultural enrichment, but also by acting as a safety valve—an alternative. People today can live in small communities because they choose to, not as in earlier times because they are trapped.

The social argument for smallness is beguiling because we know from daily experience that small is beautiful for dinner parties and, on a more serious level, those associations of love and friendship that are so important to us all. But it does not follow that we must opt for a nation of small farms and villages, nor for a way of life that relies upon small (so-called appropriate) technologies. This vision is not so much Utopianism as it is a recoil from the vitality of civilization.

It is argued that in large communities, dependent upon large technologies, the average citizen loses his or her sense of self-reliance and thus becomes *alienated*. This, as I see it, is the heart of the small-is-beautiful argument. It is true that large technology makes specialists of us all–or, to put it the other way–specialization is what makes large technology possible. Yet, in what way is specialization an evil? Self-reliance is admirable: we seek it for ourselves and teach it to our children. It feels good to cook a meal or paint a house. The making of hand-crafted furniture is, in some ways and to some people, more satisfying than being a teller in a bank. But more important than self-reliance–much more important, in my view–is the *mutual* reliance that is the basis of civilization.

I am a builder. Rely upon me: I will build you a sound building. I rely upon others–doctors, bakers, manufacturers, telephone operators, artists, politicians and, most of all, farmers. I know nothing about farming. Unless the trucks bring produce into the markets near where I live, my family and I will starve. But we have a pact, the farmers, the truckers, and roadbuilders, and I– and all the other people in this world with special talents and special needs. My ignorance of farming does not make me feel alienated, and I do not expect that a farmer or a factory worker or a typist will feel alienated just because he or she does not understand, or wish to understand, high technology. Through our willingness to rely upon each other, we have created high technology, and at the same time created the glories of high civilization. In doing so, we have lost neither our dignity nor control of our own destiny–in fact, we have gained more dignity and more control of our destinies–except in comparison to fabled arcadias that never were.

When people saw the first photographs of the earth as viewed from space, there was an outpouring of reverent philosophy. The planet seems so small, it was said, and we can now see that we are all members of a global family. Yet some of the individuals who were most intoxicated with this idea are now found among the ranks of the small-is-beautiful school, bewailing the fact that large social organizations are alienating.

I am bemused when I hear such people complain about the complexity of our contemporary technological society, and then wax rhapsodic about ages past when people purportedly had control over their own lives. An illusion of control is what they had, at best. The planet was always small, and when John Donne said that no man is an island he was stating an ecological fact as well as a poetic insight.

Picture a man of some earlier time, plowing his own field, apparently working out his own destiny. Then realize that just over the horizon locusts were approaching, or plague, or fire, or flood, or Ghengis Khan and a hundred thousand murderous swordsmen. Was that man more fortunate than the citizen of a modern Western democracy? Was that man more in command of his own life than the contemporary Everyman who works in a factory, has trouble paying his bills, and complains about his Congressman?

History tells us of advancing deserts, disappearing supplies of wild game, diminishing water supplies, and silting harbors, all of which led to a brutish struggle for survival and perpetual wars. Only with the coming of large technological systems and large social organizations have we had the chance to gain some measure of control over our fates. We seek to preserve the wholesome features of tribal life—in ways that range from PTA meetings to church socials—even as we join together in extended associations. But to shrink from organization on a large scale because of the yearning for a mythical Eden, is to breed feelings of helplessness, not to avert them. The social argument on behalf of smallness is no more coherent than the technical and political.

Perhaps the new worship of smallness is partly an aesthetic revulsion. Admittedly, American culture has much in it that is big and brash. But an association of bigness with lack of taste is simply not warranted. The colossal works of humanity are no more inherently vulgar than the small-sized works are inherently petty. We seek robustness in life as well as delicacy. Jean-Jacques Rousseau, coming upon an enormous Roman aqueduct, remarked that he felt, along with the sense of his own littleness, something which seemed to elevate his soul. To be incapable of feeling that elevation of soul is to be diminished as a human being.

"Man is small," says Schumacher, "and, therefore, small is beautiful." The statement cannot withstand a moment's thought. By what standard is man small? As human beings we are able to say that a mouse is small and a mosquito even smaller; but what does this signify? Does this make them more beautiful than a tiger or a horse? Is a stream more beautiful than the Grand Canyon? Is a manuscript illumination more beautiful than the ceiling of the Sistine Chapel? But why go on? The aesthetic argument is so absurd that we must assume that "small is beautiful" is not intended to be taken literally.

One other argument remains. Even if the small-is-beautiful doctrine does not hold together in any logical sense, it might still be founded on a base of moral conviction. Schumacher and Lovins and their followers tell us that there are qualities that we lack, and toward which we should be striving. These are thrift, simplicity, and humility. We have sinned by being wasteful, ostentatious, and arrogant. Thus smallness becomes a symbol of virtue.

But when we speak of morality, we have to be cautious. Often the most worthy moral precepts have a dark underside; take, for example, patriotism, in the name of which many a deplorable deed has been performed. In the small-is-beautiful debate, the thrift and humility being preached evoke those Oriental attitudes that perpetuate misery by teaching the masses to accept fatalistically their wretched lot. The lessons of Oriental philosophy, so useful in adding a measure of serenity to our personal lives, are insidious elements to inject into debates on public policy.

Large technology—at least as it has developed in the Western democracies—tends to promote the physical well-being of all citizens, and in this respect I would call it a towering force for moral good. If we abandon this technology before the needs of impoverished masses are attended to, I say that we will be guilty of unethical behavior. And if we turn against this technology because it has failed to make us all happy, then I fear that we are saying less about technology than we are about our own lack of maturity.

Finally, for those people who prize the arts and find moral value in their advancement, it is worth remembering that high culture goes hand in hand with high technology. Shortly after hearing E. F. Schumacher deliver a lecture extolling the virtues of smallness, I happened to attend a performance of *Der Rosenkavalier* at the Metropolitan Opera House in New York City. As the curtain rose at the beginning of Act II, the audience gasped at the sight of the blazingly illuminated reception hall of an eighteenth-century Viennese mansion. Octavian entered, gleaming in snowy damask and handed the silver rose to Sophie, who started to sing that incredibly lovely melody ("It seems like a flower from heaven, one of the roses of Paradise. Where can I have felt such joy before?"). At that moment I was convinced that if Mr. Schumacher had been sitting next to me, he would have undergone an instant conversion, and that no more would have been heard about the pleasures of primitive village life.

If that experience sounds overly genteel, let me refer instead to a 1979 NBC telecast, "Live from Studio 8H." It featured soprano Leontyne Price, violinist Itzhak Perlman, and conductor Zubin Mehta leading the New York Philharmonic Orchestra. I call attention to the event not only because of the tens of millions of people who were able to enjoy an hour of sublime music, but also because of the artists themselves: a black woman born in Laurel, Mississippi, and educated in Ohio and New York; a young man born and raised in Tel Aviv, who cannot walk without crutches; and an Indian born in Bombay, sent to Vienna to study music, and appointed leader of several great orchestras after winning first prize in a conductors' competition in Liverpool, England. What would have become of these individuals in a world that shunned "bigness"?

I have said that the small-is-beautiful movement is potentially dangerous, and the remark requires some amplification. Schumacher's book *Small Is Beautiful* has sold about a million copies. During his 1977 lecture tour in the United States he was heard by 60,000 people, and he met with the President in the White House. Amory Lovins has also reached a large audience, including people in high places. Their disciples are everywhere. *The Whole Earth Catalog,* which expounds the do-it-yourself philosophy of the counterculture, has sold two million copies and won a National Book Award. The Department of Energy has an Office of Small Scale Technology, and the Department of the Interior has established an Appropriate Technology Division. This is good to the extent that such initiatives are constructive and supplement our other social and technical endeavors.

But if such activities breed an unreasoned hostility toward large-scale enterprise, the results may be disastrous. We dare not turn our backs on macro-engineering just when we need it most, nor shun the large social organizations that are our best hope for peace and worldwide commerce. If we do, then we will end up not in halcyon villages, but back in the caves where we began.

In *Gulliver's Travels,* you may recall, the empires of Lilliput and Blefescu went to war over the issue of whether eggs should be broken at the big or little end. One of the Lilliputians explained the circumstances to Gulliver in the following manner:

> The primitive way of breaking eggs before we eat them, was upon the larger end: but his present Majesty's Grandfather, while he was a boy, going to eat an egg, and breaking it according to the ancient practice, happened to cut one of his

fingers. Whereupon the Emperor his father, published an edict, commanding all his subjects, upon great penalties, to break the smaller end of their eggs. The people so highly resented this law that, our histories tell us, there have been six rebellions raised on that account. . . . It is computed that eleven thousand persons have, at several times, suffered death, rather than submit to break their eggs at the smaller end.

Much of today's debate over big versus small is reminiscent of this fictional dispute. *Smallness,* after all, is a word that is neutral–technologically, politically, socially, aesthetically, and, of course, morally. Its use as a symbol of goodness would be one more entertaining example of human folly were it not for the disturbing consequences of the arguments advanced in its cause.

# 9

# On-the-Job
# Enrichment

A collection of E. F. Schumacher's speeches, published posthumously in 1979, was called *Good Work*. This was an appropriate title, since the importance of work—the quest for fulfillment, or even salvation, in work—is a topic that, more than any other, roused Schumacher to higher levels of passion. Industrial society, according to Schumacher, makes most forms of work "utterly uninteresting and meaningless." This is because "mechanical, artificial, divorced from nature, utilizing only the smallest part of man's potential capabilities, it sentences the great majority of workers to spending their working lives in a way which contains no worthy challenge, no stimulus to self-perfection, no chance of development, no element of Beauty, Truth, or Goodness."

At the same time, and in the same speech, Schumacher deplores the complexity of work in a modern society, and yearns for the simple physical tasks of bygone days:

> It is obviously much easier for a hard-working peasant to keep his mind attuned to the divine than for a strained office worker.
>
> I say, therefore, that it is a great evil—perhaps the greatest evil—of modern industrial society that, through its immensely involved nature, it imposes an undue nervous strain and absorbs an undue proportion of man's attention.[25]

The inconsistency here is breathtaking. We are urged to aspire to work that contains challenge, stimulus, and chance of development, while at the same time seeking simple, routine tasks that will free our mind for spiritual contemplation. It never seemed to occur to Schumacher that he was confronting an elemental enigma of human existence, and trying to make it fit within the confines of his simplistic attack on modern technology.

Questions surrounding work and its discontents are as old as civilization. During the 1970s, however, these questions took on renewed urgency, and appeared as a key element in the anti-technology campaign. Schumacher was only one of many observers to express concern about conditions in the workplace. While he was attempting to prescribe a cure for worker dissatisfaction through a return to the fields, a number of sociologists and industrial psychologists sought solutions in a new endeavor which they called "job enrichment" (or, alternatively, "work reform," "job redesign," "humanization of work," or "the quality of work movement"). The deplorable effects of technology, it was hoped, might be mitigated if humanistic concepts could be introduced into the design of work, an area traditionally dominated by "unfeeling" engineers and business managers.

In early 1972 workers at the Vega plant in Lordstown, Ohio, went out on strike, not for more money or shorter hours, but to protest the pressure and monotony of their work on General Motors' fastest-moving assembly line. That 23-day work stoppage helped make "worker alienation" a fashionable term in industrial, sociological, and literary circles.

At the time of the Lordstown strike, a stream of reports were arriving from Sweden, where SAAB and Volvo were trying to deal with worker discontent by experimenting with alternatives to the assembly line. These companies attempted to give workers a sense of significance by having them participate democratically in decisions affecting the manufacturing process. The initial successes attributed to these efforts were reported in a series of beguiling newspaper and magazine articles.

At year's end the Department of Health, Education and Welfare released a study, *Work in America,* which reported that people at all levels of society were becoming increasingly dissatisfied with the quality of their working lives, to the detriment of the economic and social well-being of the nation. This study, widely distributed and acclaimed, provided a manifesto for the revolution that Lordstown seemed to portend. With alacrity the concept of job enrichment spread through the worlds of journalism, academe, business, and government. Corporations hastened to establish ways by which workers could help make decisions affecting their jobs. In *The Future of the Workplace,* completed at the end of 1974, Paul Dickson concluded that newly devised "humanization of work" experiments were proving so successful that corporate executives were beginning to view them as important proprietary developments whose details were not to be shared with competitors. Dickson reported that these changes were no passing fad, but harbingers of things to come.

Concern for the alienated worker, in addition to spawning a host of industrial experiments, articles, studies, grants, and conferences, also inspired Studs Terkel's bestseller *Working* and Barbara Garson's *All the Livelong Day: The Meaning and Demeaning of Routine Work*. Both books were based upon interviews with workers and, in the words of the people themselves, the message seemed to be unambiguous: Americans hated their jobs. They left them frustrated and demoralized. Americans seek in their daily occupations a sense of identity, self-esteem, autonomy, and accomplishment. What they get, according to Terkel, is "daily humiliations." Their fragmented, monotonous jobs are, in Garson's view, "soul-destroying." The average worker's discontent manifests itself in fighting, swearing, absenteeism, high turnover rates, sabotage, alcoholism, drug addiction, and poor mental health. Reform of the workplace, it seemed, was one of the most critical social issues of our time.

Nevertheless, just as enthusiasm for work humanization was reaching a fever pitch among intellectuals, disenchantment set in at many of the places where the experiments were taking place. In a September–October 1975 *Harvard Business Review* article, J. Richard Hackman, an organizational psychologist, reported that "job enrichment seems to be failing at least as often as it is succeeding." And further: "Even though the failures may be relatively unobtrusive now, they may soon become overwhelming."

Corporate executives were not the only ones disappointed with the results; among the workers and union leaders interest also appeared to be flagging. The United Auto Workers, for example, appeared to have forgotten about Lordstown. In preparing for new contract negotiations, they were concentrating on the issues of wages and job security.

An obvious conclusion was that a recession was occurring just in time for management to put the rebellious workers in their place. Clearly, in uncertain times, most people are less interested in fulfillment than in a living wage. But the supporters of job enrichment claimed that a more satisfying job results in improved productivity, so that it should be a management goal in bad times as well as good.

In fact, it was this very feature that had aroused the suspicions of labor union leaders whose lack of cooperation appeared to be one of the main reasons for the many failed experiments. Job enrichment, according to a vice-president of the International Association of Machinists, is "a speed-up in the guise of concern for workers." Any experiment that results in increased productivity is necessarily suspect. Even assuming the best of motives, it is disturbing to note that job enrichment depends upon manipulation of workers by the experts. In this respect, it can be viewed as an extension of the much-maligned art of scientific management. The experts, of course, maintain that the new redesign of work is done in response to the desires of the workers. Yes, but it is the experts who must determine what these desires are—a subtle and troubling point.

What *do* people want out of life? That is one of those questions whose answer can be shaped by the way in which the question is posed. Straightforward statistical studies find that in apparent contradiction to Terkel's findings, job discontent is not high on the list of American social problems. When the Gallup Poll's researchers ask, "Is your work interesting?" they get 80 to 90 percent positive responses. But when researchers begin to ask more sophisticated questions, such as, "What type of work would you try to get into if you could start all over again?" complaints begin to pour forth. The probing question cannot help but elicit a plaintive answer. Which of us, confronted with a sympathetic organizational psychologist, or talking into Studs Terkel's tape recorder, could resist tinging our life's story with lamentation, particularly if that was what the questioner was looking for? Compared to the labor performed by most people in the past, today's jobs seem quite attractive. Compared to the "calling" that Terkel says we are all seeking, what job could measure up?

Indeed, people are not "satisfied" with their work nor with any other aspect of their lives. This is hardly news. But can we agree on what should be done to improve the situation? Barbara Garson sees a solution only in workers controlling their own jobs through socialism. (The widespread dissatisfaction of workers in socialist countries does not impress her.) Most proponents of job enrichment, while not advocating socialism, agree that what the average worker misses most is a sense of responsiblity and participation in the making of decisions. But is this assumption valid? Are there not many workers who do *not* want responsibility, who prefer the comfortable monotony of routine tasks to the pressures of making decisions and being accountable for the consequences? Miss Garson's workers keep contradicting her basic premise. From a woman who has turned down the job of supervisor: "I don't need the responsibility. After work I like to spend my time fixing up my house. And that's what I like to think about while I'm working." And from people with mechanical, repetitive tasks: "Flip, flip, flip . . . feels good," "you can get a good rhythm going," "you kind of get used to it." Even Garson despairs for a moment: "Maybe the reactionaries are right. Maybe some people are made for this work."

To have thought so (or to have admitted it) up until recently would indeed have marked one as a reactionary. But times are changing. The work enrichment movement appears to be running counter to another trend, the seeking of inner peace rather than ego fulfillment. In the light of this new wisdom, which advocates, among other things, the blanking of the mind in meditation for an hour each day, one can wonder who has the better of the bargain, those who are in the ratrace or those who are "beneath" it. This is the paradox that leads Schumacher into his capricious inconsistencies.

In one episode, Garson tells of a small commune in which ten young adults lived on the wages of four, and where the focus of life was away from work. I dare not predict what modes of life will be attractive to the masses of the future. I believe, however, that the job enrichment enthusiasts have made a mistake in assuming that all people desire what social scientists want them to desire. From Lordstown and some amorphous complaints they have made unwarranted extrapolations.

An even more glaring mistake is that of assuming that by restructuring the workplace, one can solve the problem of alienation. This hypothesis calls to mind those urban planners who saw salvation for the poor in a clean, spacious apartment, and who, after their ideas have been carried out, have spent much energy explaining why attractive apartments have not, in fact, eradicated the ill effects of poverty. They wander from failure to failure seeking the magic environment (high-rise, low-rise, slum clearance, renovation, vestpocket projects, town houses) like so many Ponce de Leons trudging through the malarial Florida swamps.

Alienation cannot be cured by a fascinating job any more than it can be cured by a clean apartment. Some of the best jobs, by almost any standard, are held by members of the skilled construction trades. These people do interesting, varied work. They are craftsmen in the tradition that Schumacher admires. They are not too closely supervised. They see the tangible results of their labor. Their strong unions have made sure that they do not have to produce more than they can comfortably handle. They are well paid. E. E. LeMasters spent five years mingling with hardhats in a tavern, and reported on his experience in a book entitled *Blue-Collar Aristocrats*. He found that these men are pleased with their work and are proud of what they do. He also found that they are about as alienated as it is possible to be – alienated from their wives, their children, their churches, and their political leaders. They are bigoted and full of hate, confused and full of suspicion.

There are diseases of the soul abroad in the land, but only a few of the symptoms, not the viruses themselves, are to be found in the workplace. Healthy people do not become heartless bosses or cruel foremen. Healthy people do not feel debased or dehumanized by menial work or intimidated by blustering superiors. Sick people–alienated people–are not made whole by an interesting job.

Of course, the concept of job enrichment has much to commend it. The idea that work should provide satisfaction is worthy of further pursuit. It serves the interests of the workers, as long as they are assured of not being subtly manipulated, and ideally, it serves the interests of industry and all society, by resulting in increased productivity.

The Japanese seem to have been uniquely successful in this endeavor, indicating that worker alienation has less to do with industrialization than it does with other aspects of the general culture. Work in a factory or a large office is not inherently less satisfying than work on a farm or in a small-town store. The "dehumanization of the workplace" is only tangentially related to technological advance. It is mainly attributable to the way people feel about themselves, and the way that they treat each other. We do not have to look to the Japanese for proof. Anybody who has worked knows that this is so.

In identifying industrial work as a major source of contemporary malaise, the antitechnologists divert us from asking ourselves what we can do to improve mental health, foster common courtesy, and nourish concern of one person for another.

Those who would blame all of life's problems on an amorphous technology, inevitably reject the concept of individual responsibility. This is not humanism. It is a perversion of the humanistic impulse.

# 10

# Codifying the Future

The 1970s could well be described as the golden age of regulation. During that decade, not only were new laws passed and new federal agencies created—the Environmental Protection Agency, the Council on Environmental Quality, the Occupational Safety and Health Administration, and the Consumer Product Safety Commission—but also the older regulatory agencies were revitalized. Spurred by new rulings, new court decisions and, most of all, by new public attitudes, the older agencies launched a multitude of regulatory initiatives. ¶Of course, regulation, properly defined, is not hostile to technology. A complex technological society requires an abundance of regulation; indeed, regulation is itself technological. But irrational zeal for excessive regulation is quite another matter. This represents antitechnology in its most virulent form.

The regulatory impulse, running wild, wreaks havoc, first of all by stifling creative and productive forces that are vital to national survival. But it does harm also—and perhaps more ominously—by fomenting a counter-revolution among outraged industrialists, the intensity of which threatens to sweep away many of the very regulations we most need.

I become disturbed when I see an illustration in *Exxon USA* portraying the State of Liberty enveloped in red tape, or advertisements from the Amway Corporation showing an ugly "Federal Nanny" hovering overhead or a "regulatory" branch choking the other branches of the tree of liberty. If regulations are often imperfect, or even absurd, that is because regulators are as fallible as the bankers and department-store executives who regularly foul up our personal accounts. Regulations need to be rationalized, of course, but not disparaged and weakened to the point that apathy is tolerated and avarice given free play. Bureaucracy is the price we pay for technological complexity and creative greed. The Statue of Liberty is not tied up in red tape, dear Exxon—she is held together by red tape.

If we do not check the relentless, almost mindless, efforts of some antitechnologists in the regulatory arena, the 1980s may come to be known as the decade in which, after first weakening our capacity to produce, we then, in disgust, sacrificed essential environmental safeguards.

There are many examples that might be used to demonstrate the type of self-defeating activity I have in mind, but there is one that I consider to be particularly apt. It is the effort of the Federal Trade Commission to enact a trade-regulation rule governing the voluntary organizations that establish industrial standards.

In Washington hearing rooms, as I have seen them on television, dramatic events unfold before tense audiences. But one morning in late September 1979, when I entered the third-floor hearing room in the Federal Trade Commission building, the participants in the proceedings then underway seemed overcome by torpor, and no audience at all was there to witness the end of long hearings on a proposed trade-regulation rule covering "product standards and certification." Starting in San Francisco and concluding in Washington, for 12 weeks, a small group of specialists had been arguing about certain arcana of our industrial society: the procedures by which volunteer experts establish thousands of technical standards for materials and manufactured products.

During much of the morning only seven persons, besides me, were in the room: the FTC-appointed presiding officer, a witness from the Institute of Electrical and Electronics Engineers, his lawyer, a stenographer and, to examine the witness, one representative each from the FTC, the National Consumers League, and the American National Standards Institute (ANSI), the umbrella organization for most of the standards-setters in the nation.

Despite the hush and the evident boredom, I felt that I was present at an occasion of some importance. I sensed that I was witnessing a crucial defeat for the forces of government regulation in the United States. Like a stranger coming across an advance platoon of Napoleon's army before the gates of Moscow, I thought, "The cause is lost; they have chosen the wrong enemy and come too far, recklessly; the great retreat starts here."

I first learned about the proposed FTC rule in the pages of engineering journals. *"Voluntary Standards Under Attack,"* read a headline in the December 1978 issue of the *ASCE News,* a publication of the American Society of Civil Engineers. *"Federal Trade Commission Reaches for Regulatory Input To Every Segment of Standards-Setting Operation,"* announced a feature in *Professional Engineer. "Another Incursion Into Private Enterprise,"* sputtered the editors of *Consulting Engineer.*

I found it hard to believe what I was reading. Who would dare to attack the voluntary standards-setters of America? Who could be foolish enough to challenge the 300,000 individuals who volunteer some of their time, under the auspices of trade, technical, and professional organizations, to writing the industrial standards of the nation? One might sooner launch an assault on the League of Women Voters. I could only conclude that someone at the FTC had gone mad.

My sense of alarm did not arise out of any personal animus toward government regulation. On the contrary, as I have said, it was out of concern for the beleaguered cause of regulation that I deplored the move against the standards-setters. I even fancied that the public relations department of the U. S. Chamber of Commerce had planted a double agent within the FTC whose mission was to launch an assault on a placid yet powerful community whose history is one of the triumphs of American democracy.

Although standards are as old as civilization (in 1266 Henry III of England decreed that a penny was to weigh the equivalent of 32 grains of wheat "taken from the middle of the ear"), the modern age of standards began in the nineteenth century with the development of mass production. If manufactured parts were to fit together and be interchangeable, and if parts made in one factory were to be assembled in another, then there had to be agreement on dimensions and quality of materials—that is, there had to be standards. Such simple and ubiquitous items as the nut and bolt were being made haphazardly in thousands of sizes, shapes, and screw-thread configurations, a situation that an emerging industrial society could not tolerate. Wherever people of science and industry gathered, the need for standardization was discussed. At a meeting of the Franklin Institute in Philadelphia in 1864, a Mr. William Sellers proposed a system for standardizing screw threads that within a few years gained wide acceptance. The American Society of Civil Engineers established a committee to develop a standard steel rail; the American Society of Mechanical Engineers set to work on a code for steam boilers. In 1898, a nonprofit organization called the American Society for Testing and Materials started to codify standard sizes, strengths, and other characteristics for the burgeoning steel industry. Intercompany standards associations were sponsored by several industries, notably the railroads, which required prototypical equipment such as safety couplings and air brakes, to say nothing of a standard track gauge to replace the 33 different dimensions that were in use at one time.

As the need for standards outstripped the facilities to provide them, the National Academy of Sciences pressed Congress to establish a national standardizing laboratory. In 1901 the National Bureau of Standards was founded, modeled after Germany's Imperial Physical-Technical Institute (organized in 1887) and England's National Physical Laboratory (established in 1900). In addition to taking over and expanding the Treasury Department's Office of Weights and Measures, the NBS was given the responsibility of making tests to guide the purchases for federal departments. It thus became a technical resource for both industry and government, researching, testing, and setting standards for myriad materials and products–cement, light bulbs, paper, twine, resins, varnishes, and so forth. By 1911 the bureau was conducting some 80,000 tests annually, and sending inspectors into every state to examine the scales of shopkeepers (most of whom were found to be shortweighting their customers). Industry, however, did not want all standards to become the province of a federal agency, and even NBS officials agreed that such an assignment would swamp them in petty details and subject them to unwanted political pressures. So the idea of making all standards a responsibility of the federal government, although advocated by some, was not implemented.

As dozens of corporations, trade associations, and professional societies increased their standardizing activities, overlapping and conflict inevitably occurred. Thus, in 1916, the professional societies of the civil, electrical, mechanical, and mining engineers, along with the American Society for Testing and Materials, met to discuss the coordination of standards on a national level. After two years of conferences, these five societies established the American

Engineering Standards Committee. The purpose of this organization was not to create standards, but to review and coordinate those being developed elsewhere. The Departments of Commerce, War, and the Navy accepted invitations to become founding members. Soon other government agencies, and then many trade associations, joined, and in 1928 the committee reorganized and changed its name to the American Standards Association. (In 1948, when the association incorporated under the laws of New York State, federal agencies withdrew from formal membership, although their personnel remained active on technical committees. The present name, the American National Standards Institute, dates from 1969.)

Although the standards movement was occasioned by mass production, it made its way into many areas of American life. Insurance companies, concerned about fire losses and electric shock hazards, founded Underwriters Laboratories in 1894 and, two years later, the National Fire Protection Association. A growing sensitivity to the rights of workers was expressed in a drive for industrial-safety codes, launched in 1919 at a meeting that standards experts held with representatives of labor, industry, and government. Building codes were developed, and standards adopted for pharmaceuticals and agricultural products. The consumer movement was created in the late 1920s, not by activist lawyers, but by standards professionals. During Herbert Hoover's term as Secretary of Commerce (1921–29) a "Crusade for Standardization" became popular. The number of mattress sizes was reduced from 78 to four, varieties of milk bottles from 49 to nine. Satirists predicted that eventually standardization would reach ladies' hats.

Today there are more than 400 private organizations—trade, technical, professional, consumer, and labor—that have written or sponsored the approximately 20,000 current commercial standards. Most of the larger of these organizations are members of ANSI, which defines itself as "the coordinating organization for America's federated national standards system."

By far the most prolific member of this community is the 83-year-old American Society for Testing and Materials. Almost 30,000 individuals serve without pay on the ASTM's 135 standards committees. ASTM standards are usually developed at the request of an industrial trade association or a government agency, and they come in many varieties: a specification for stainless steel bar and wire for surgical implants; a recommended practice for rating water-emulsion floor polishes; a method of making and curing concrete test specimens in the field; a classification system for carbon blacks used in rubber products; a method of testing tires for wet traction in straight-ahead braking, using conventional highway vehicles.

To assure committee balance, the ASTM requires that neither the chairman nor more than half the members can be "producers." Draft documents prepared by task groups are reviewed by a mail-balloting procedure and, finally, by the entire ASTM membership. At each point along the way, negative ballots accompanied by written comments must be considered; dissatisfied voters can appeal to the board of directors committee that grants final approval. The ASTM standard is usually submitted to ANSI for endorsement as an American National Standard, and another routine begins, one that in recent years has given particular attention to the interests of small businesses and consumers. Many published standards find their way into government specifications.

Other leading developers of standards in the ANSI system are the Society of Automotive Engineers, the American Society of Mechanical Engineers, the Institute of Electrical and Electronics Engineers, Underwriters Laboratories, the American Petroleum Institute, the Electronic Industries Association, and the National Fire Protection Association. Although not all of these organizations follow the full-consensus procedures of the ASTM, nor seek the participation of nontechnical people, their members all do view themselves as conscientious professionals.

Anyone who reviews the history of voluntary standardization in the United States cannot fail to be impressed by the benefits that arise from the activities of this unique social institution. And anyone who reads in the literature of the standards-setters themselves cannot help noting how proud they are of what they do. This is the community that the FTC staff proposed to subject to stringent, wide-ranging, and unprecedented regulatory control.

During the session I attended, the principal witness was Ivan G. Easton, consulting director of standards for the Institute of Electrical and Electronics Engineers, whose calm monotone did not conceal his annoyance. "We are dealing with high technology," he said, speaking of the IEEE's standards-making activities. There is no need for a new FTC rule, Easton argued, one that would only open the door to trivial challenge and harassment from people whose main interests lie outside the standards field. Instead of being pestered by lay bureaucrats, Mr. Easton implied, he and his colleagues should be thanked and encouraged to continue their constructive work.

The young FTC lawyer who was doing the questioning, however, saw Mr. Easton in a different light. "Aren't the people on your committees sponsored by their employers?" he asked. "And is this totally altruistic?" The implication that the IEEE's standards committees are controlled by the large corporations is wildly ironic, particularly in light of the fact that the 160,000-member organization has long been considered the most radical of the engineering societies, implacably opposed to domination by business interests. It went so far as to resign from the Engineers Joint Council when that organization in 1967 decided to accept corporate members.

After the lunchtime break, in the hope of discovering the rationale for what increasingly appeared to me to be a lunatic proceeding, I sought out the responsible members of the FTC staff, and was directed to the office of Robert J. Schroeder, project manager. I yield to no one in deploring prejudice against the young, but I did find it unsettling to learn that Mr. Schroeder, five years out of the University of Michigan Law School, and four other young men of approximately the same experience constituted the entire legal staff of the FTC's Bureau of Consumer Protection, the body that was proposing to reform one of the nation's venerable technical institutions.

It all began in 1974, Mr. Schroeder explained, with the case involving foamed-plastics insulation. Manufacturers had been marketing plastic insulation as "nonburning" and "self-extinguishing," using as justification an ASTM test that exposed a small piece of plastic to an open flame. However, when buildings burned, plastic insulation burned along with them, giving off a poisonous smoke. After a number of deaths were attributed to the flammability of foamed plastics, the FTC issued complaints against 25 manufacturers and their trade association, and named the ASTM as "the means and instrumentality" involved. Under a consent order, the manufacturers and their association agreed to stop making the unwarranted claims, to notify past purchasers of the danger, and to conduct a $5 million research program.

The ASTM, maintaining that its test results had been misused by others, did not participate in the consent order, and when the FTC did not persist, declared itself vindicated. The FTC, however, asserted that the plastics industry had "captured" the ASTM committee, and used it to issue a self-serving standard. It is because of this, Mr. Schroeder said, that the FTC commissioners recommended a general investigation of the standards field.

Four years of study convinced the FTC staff that a rule-making procedure was justified. Standards are called "voluntary," Mr. Schroeder explained, but once they are adopted for use they are likely to become mandatory, and are often given the force of law. Therein lies a potential for abuse. This is recognized by the standards-setting organizations, which strive to avoid it by balancing their committees and establishing other democratic procedures. "But we have seen injury to consumers and to competitors," Mr.

Schroeder insisted. "It's okay to say that the system works, but just read some of the instances that we have uncovered."

On my return home I did read through the FTC staff's 572-page report, and found about 30 instances of purported abuse of the standards process. There were several examples of what the report calls "buyer misreliance," of which the most prominent was the flammable-plastics case. A few of the others:

- Aluminum electrical wire, after being approved by the Underwriters Laboratories in the 1960s, was implicated as a fire hazard and found to require special connecting devices. The UL, it is claimed, was slow to modify its standards and approval practices.
- Lighting-level standards developed by the Illuminating Engineering Society were steadily increased over the years. With the coming of the energy crisis, they were deemed to be wastefully high and were belatedly decreased. (This has been a special grievance of Ralph Nader.)
- An ASTM standard for brick was criticized for failing to state a minimum initial rate of absorption (of water from fresh mortar), which might affect the ultimate strength of a brick wall.

More numerous than "buyer misreliance" complaints were instances of purported "product exclusion." For example:

- Plastic pipe was kept out of plumbing codes long after it was found suitable for certain uses. Presumably this happened because of pressure from plumbers' organizations, which preferred the more labor-intensive cast-iron pipe.
- It took ANSI six years to develop a standard that permitted toughened glass as an alternative to porcelain for use in high-voltage electrical insulators.
- A manufacturer of loose-fill powder insulation for under-

ground pipe complained that his material was unfairly excluded from a standard developed by the Building Research Advisory Board of the National Academy of Sciences.

• When the Railroad Uniform Freight Classification Committee approved the use of foamed-plastic packaging, the manufacturers of traditional cellulose packaging protested that their material outperformed the new product.

Other complaints were filed by manufacturers of boiler low-water cutoff devices, relief valves for hot-water heaters, screw-thread gauges, safety spectacles, inexpensive sprinkler systems (not approved by the National Fire Protection Agency), butt-welded intermediate metallic conduit (banned by Underwriters Laboratories), thin ceramic tile, water system backflow prevention devices, automatic vent dampers for gas furnaces, burglar alarms, bathtubs, and wine bottles.

In this catchall of complaints, it is difficult to determine which values the FTC staff meant to espouse. It opposed hasty approval of new materials, as well as over-long deliberation or footdragging. It deplored economy at the expense of safety, and safety at the expense of economy. It condemned practical compromise (noting with disapproval that "decisions are susceptible to being based more on political give-and-take among various factions than on sound technical/evidentiary grounds"). It also condemned the "mistaken assumption" that there are any "unbiased experts." It was, of course, against errors and in favor of perfection.

In reading the report I could only wonder at the complexity of the issues with which the standards-setters contend, and marvel at the way in which so many interests seem to be accommodated. The processing of more than 20,000 standards had resulted in fewer than 100 dissatisfied parties (including those who were heard during the course of the hearings), and many of these were soreheads who in no way discredited the people they criticized. What other institution, public or private, has done as well?

The report admitted that those who feel aggrieved by the standards system do have recourse, first making use of ANSI procedures, then appealing to the media or Congressional committees, and finally filing private antitrust and products-liability actions. (Only 30 formal complaints had been filed with ANSI in the ten years prior to the report and all of these had been resolved without litigation.) The report conceded that in the past the FTC has dealt with standards problems by issuing industry guides and advisory opinions (in 1970, in response to a request from ANSI itself). It admitted that the standards organizations have taken steps to update and improve their procedures. I looked through the report in vain for a clue to how society might benefit from the mind-numbing regulatory document that had been produced.

The Proposed Trade Regulation Rule for Standards and Certification covered 16 pages of tightly packed type. ("This isn't all bad," an ANSI official confided to me, "the standards organizations' lawyers love it.") The rule required that "notice" be given to the public at three different stages of a standard's development; it required that all persons (including environmental groups and energy–conservation groups) have equal opportunity to participate in all phases of standards proceedings; it contained a provision for "duty to act" in response to any legitimate challenge, and a definition of "appropriate action" (withdraw the standard, revise, or cease to distribute). There were sections on "required disclosures," and "record-keeping and access," followed by a section on "appeals" (each standards-setter would have to establish an independent appeal board). Finally, there were sections dealing with the special responsibilities of certifiers and marketers (the information required to be included on labels would sometimes make the label larger than the product).

While ANSI did not quarrel with the rule's main objective—a fair representation of all interests in the standards process—it contended that the new regulations would mean substantial added administrative cost for the larger standards organizations, and would probably drive smaller standards-setters out of the business altogether. Obstructionists would have a field day, and the existing cadre of talented volunteers would become dispirited. And to what end? "Opponents of the FTC rule," said ANSI's official statement, "are being forced to defend a fantastically productive and effective standards system, which is the envy of the world."

Moreover, ANSI's lawyers maintained that the proposed rule imposed "prior restraint" on ANSI's right to publish standards, and so constituted a violation of First Amendment rights. They claimed further that the FTC would not have authority to implement the rule, since ANSI, as a nonprofit corporation, is not subject to the FTC's jurisdiction.

Finally, the standards organizations pointed out that existing laws are adequate to remedy flaws in the system. The FTC is *already* empowered to prevent "unfair or deceptive acts or practices." In addition, the Sherman Act is enforceable by the Justice Department, the Consumer Product Safety Act by the Consumer Product Safety Commission.

The FTC lawyers said that they were merely trying to clarify laws that already exist, to better define what is "unfair" and what is not. ANSI's reply was that if the law is to be modified, then it is up to Congress to do it, and that the Senate Antitrust and Monopoly Subcommittee, after holding hearings in 1975, 1976, and 1977 on purported standards abuses, decided that no new legislation was warranted.

The hearings ended the day after my visit. Ralph Nader, who had been scheduled to testify, canceled his appearance, a good indication that the rule's advocates considered it a lost cause, at least in the form proposed. ("We knew that he wouldn't show," ANSI's counsel chuckled.)

During early 1980 there developed in Congress a revolt against the FTC that came close to stripping that agency of many of its long-established powers. Much of this initiative stemmed from lobbyists who resented the FTC's intrusion into the affairs of funeral directors, used-car dealers, and producers of children's television programs. But many responsible members of the scientific community, who ordinarily would have been defenders of the FTC, joined its assailants because of their indignation over the standards rule-making effort.

The Congressional attack against the FTC was eventually repulsed, mainly through the threat of a Presidential veto, but the rule governing standards-setters was held in abeyance during the turmoil. Regardless of what the future may hold, it is clear that the FTC suffered a grievous self-inflicted wound, and antagonized tens of thousands of independent-minded professionals, precisely at a moment when it could least afford to do so.

Why, then, did the FTC investigators embark on this ill-conceived venture? Some people think that bureaucrats lust insatiably for power, but I do not see that as a supportable argument. Perhaps some regulators believe that it is their role to extend their influence as far as possible on the assumption that unrelenting enemies of regulation are doing as much themselves. Such an unfocused aggressiveness, however, does not adequately explain the actions of the FTC staff.

It appears to me that Mr. Schroeder and his colleagues were motivated by a consuming impulse to codify the future. Clearly they had no interest in solving immediate problems. Those individuals who claimed to have been wronged by the standards establishment, instead of being helped, were turned into witnesses in support of some obscure future good. The Bureau of Consumer Protection was so busy making rules that the current needs of consumers went unattended.

To formulate redundant statutes instead of doing the day's work is a distortion of the regulatory function. The few complaints that arise out of the development of standards should be addressed, on a case-by-case basis, by diligent, competent investigation. But complex technological problems will not yield to another 16 pages of legalistic prose.

*Legalistic* is the operative word. It is less significant that the FTC staff is young, bureaucratic, and (let us assume) idealistic, than that they are lawyers, and thus imbued with an excessive esteem for words. Instead of using the authority they already have, they create new definitions of authority. Our shelves grow heavy with law books, and our problems go unresolved.

American society is not overregulated. It is overlegislated and undermanned, overwritten and underaccomplished. It is over-lawyered and underengineered.

Engineers tend to concentrate on the job that needs doing, and they implicitly place their faith in the ingenuity and good sense of future generations. Lawyers, accustomed to drafting contracts and executing wills, try to command posterity with the sorcery of clever phrases. The conflict over the proposed FTC rule on standards was, philosophically and literally, a battle between the two professions.

Engineers have learned in recent years that misplaced highways and parking lots can blight the lives they are intended to enhance. Let us hope that regulatory lawyers will learn, with equal grace, that the fertile fields of creativity are to be tended, and occasionally weeded, but not paved over with rules.

# 11

# The Feminist Face
# of Antitechnology

The campus of Smith College in Northampton, Massachusetts, is one of the pleasantest places in the world to be on a sunny spring afternoon. The setting is so lovely, the academic atmosphere so tranquil, that when I arrived there on such an afternoon in April, I was totally captivated. The spell of the place, however, made me uneasy about my mission, which was to convince a few of the students at this premier, all-female liberal arts college that they ought to become engineers. ¶The mission, as it turned out, was destined to fail. Most bright young women today do not want to become engineers. At first hearing this might not seem to be a matter of grave consequence, but since engineering is central to the functioning of our society, its rejection as a career option by female students raises the most profound questions about the relationship of women to technology, and about the objectives of the women's movement.

It is not generally recognized that at the same time that women are making their way into every corner of the work-world, less than 3 percent of the professional engineers in the nation are female. A generation ago this statistic would have raised no eyebrows, but today it is difficult to believe. The engineering schools, reacting to social and governmental pressures, have opened wide their gates and are zealously recruiting women. The major corporations, reacting to even more intense pressures, are offering attractive employment opportunities to nearly all women engineering graduates. According to the College Placement Council, engineering is the only field in which average starting salaries for women are higher than those for men. Tokenism is disappearing, according to the testimony of women engineers themselves. By every reasonable standard one would expect women to be attracted to the profession in large numbers. Yet less than 10 percent of 1980's 58,000 engineering degrees were awarded to women (compared to 30 percent in medicine, 28 percent in law, and 40 percent in the biological sciences). By 1984 the total may reach 15 percent, still a dismal figure when one realizes that more women than men are enrolled in American colleges. Unless this situation changes dramatically, and soon, the proportion of women engineers in practice, among more than 1.25 million males, will remain insignificant for many decades. While women are moving vigorously—assertively, demandingly—toward significant numerical representation in industry, the arts, and the other professions, they are, for reasons that are not at all clear, shying away from engineering.

At Smith I was scheduled to participate in a seminar entitled "The Role of Technology in Modern Society." The program called for a "sherry hour" before dinner, during which the speakers had an opportunity to talk informally with the students. In a stately paneled room the late-afternoon light sparkled on crystal decanters as we sipped our sherry from tiny glasses. The students with whom I conversed were as elegant as the surroundings, so poised, so *ladylike*. I found myself thinking, "These girls are not going to become engineers. It's simply not their style." The young women were not vapid in the way of country gentry. Far from it. They were alert and sensible, well-trained in mathematics and the sciences. I could imagine them donning white coats and conducting experiments in quiet laboratories. But I could not see them as engineers. It is a hopeless cause, I thought. They will not become engineers because it is "beneath" them to do so. It is a question of social class.

This was an intuitive feeling of the moment, although, when scrutinized, it made sociological sense. Traditionally, most American engineers have come from working-class families. In the words of a post-Sputnik National Science Foundation study, "Engineering has a special appeal for bright boys of lower and lower-middle-class origins." [26] Yet in many of the blue-collar families that have been such a fertile source for male engineers, the idea of a scientific education for women has not taken hold. Therefore, most of the young women who have the educational qualifications to become engineers are likely to come from the middle and upper classes. But the upper classes do not esteem a career in engineering: thus few women engineers.

We have inherited much of our class consciousness from England, and so it is with our attitude toward engineering, which the English have always considered rather a "navvy" occupation. Because engineering did not change from a craft to a profession until the mid-nineteenth century, and never shed completely its craftsman's image, it was fair game for the sneers of pretentious social arbiters. Herbert Hoover, a successful mining engineer before he became President, and something of a scholar who translated Agricola from the Latin, enjoyed telling about an English lady whom he met during the course of an Atlantic crossing. When, near the end of the voyage, Hoover told her that he was an engineer, the lady exclaimed, "Why, I thought you were a gentle-man!" The fact that this anecdote is told and retold whenever conversation turns to the role of engineers in American society indicates how basic is the point that it illustrates.

It may not be realistic to expect women to break down class barriers that were created mostly by men. Yet feminists, if they are serious in their avowed purposes, should by now have taken the lead in changing this situation, encouraging the elite among edu-cated young women to reevaluate their social prejudices. For until upper-class aversion to engineering is overcome, or until lower-class women take to studying the sciences in earnest, engineering will remain largely a male profession. And while this condition prevails, the feminist movement will be stalled, probably without even knowing it. For, in a man-made world, how can women achieve the equality they seek?

My view, needless to say, is not shared by the feminists of the United States. Judging by their literature, they seem to attach no particular importance to increasing female enrollment in engineer-ing, perhaps because they are more concerned about battering on closed doors than they are about walking through those that are open. When they do consider the problem, it is not to question or criticize choices being made by women, but only to deplore the effect of external forces.

There is an entire literature devoted to explaining how en-gineering, and to a lesser degree science and mathematics, has de-veloped a "male image." The terminology of this literature has been ringing in our ears for a long time—"sex role socialization," "undoing sex stereotypes," "self-fulfilling prophecy," and so forth. We know the facts by heart: girls learn early that it is not socially acceptable for them to play with trains and trucks. They learn from teachers that boys perform better than girls in math and science. A condition called "math anxiety" is attributed to these social pres-sures. As girls mature, they are persuaded by counselors and family that it is not feminine to enter traditionally male professions. They are afraid to compete with men or to let their intelligence show, lest they seem sexually less desirable. Finally, there is a shortage of "role models" with whom a young girl can identify.

Yes, yes, yes, of course, but these facts, which seemed so interesting and important a decade and more ago, are now stale. As the sociologists busy themselves collating their data and getting it published, the times invariably pass them by. After all, *The Feminine Mystique* was published in 1963, and the Equal Pay Act was enacted by Congress that same year. Since then a social revolution has taken place. Educated young women know well enough that they can become engineers. Surely the women who are planning to be biologists and doctors know that they could choose engineering instead, and those who are crowding into the fields of law, business, and journalism know that they could have opted for engineering if they had been willing to take a little calculus and physics. Women's magazines that once specialized in menus and sewing patterns are now overflowing with advice on how to compete in what used to be a man's world–how to dress, sit, talk, intimidate, and in general "make it." Engineering's purported male image is no longer an adequate explanation for female aversion to the profession.

It has been hypothesized that women avoid engineering because it has to do with technology, an aspect of our culture from which they recoil instinctively. Ruth Cowan, a historian at the State University of New York, has done research on the influence of technology on the self-image of the American woman.[27] The development of household appliances, for example, instead of freeing the housewife for a richer life as advertised, has helped to reduce her to the level of a maidservant whose greatest skill is consumerism. Factory jobs have attracted women to the workplace in roles they have come to dislike. Innovations affecting the most intimate aspects of women's lives, such as the baby bottle and birth-control devices, have been developed almost exclusively by men. Dependent upon technology, but removed from its sources and, paradoxically, enslaved by it, women may well have developed deep-seated resentments that persist even in those who consider themselves liberated.

If this situation does exist, we might expect that the feminists would respond to it as a challenge. The brightest and most ambitious women should be eager to bend technology, at long last, to their own will. But this is not happening. The feminists seem content to write articles assuring each other that they have the talent to fix leaky faucets.

Wherever the enemies of technology gather, women are to be found in large numbers. The transcendental movement that arose out of the counterculture of the 1960s—what Marilyn Ferguson has called *The Aquarian Conspiracy*—pits feminine sensitivity against a "macho" materialism. "Wherever the Aquarian Conspiracy is at work," writes Miss Ferguson, "women are represented in far greater numbers than they are in the establishment." This follows from basic physical and social realities: "Women are neurologically more flexible than men, and they have had cultural permission to be more intuitive, sensitive, feeling." [28]

Such an outlook not only explains why women are likely to be hostile to technology, but also raises the question of whether or not women are equipped biologically to excel in engineering. This is a theory that arouses such rancor that I hesitate to bring it up, and yet it must be confronted. The intellectual factor most closely related to achievement in science is spatial ability, the ability to manipulate objects mentally. Experiments have shown that males are, on average, better at this than females, and that this superiority appears to be related to levels of the male hormone testosterone.

It is a mistake, I think, to argue as some feminists do that there is no discernible difference between the male and female brain. It would be more sensible to say that because of substantial overlap in test scores, the differences that do exist are not practically significant when one considers a large group of potential engineers of both sexes. It would be better yet to point out that such differences as there are would serve to enrich the profession, since good engineering requires intuition and verbal imagination as well as mathematical adeptness and spatial ability. In their so-called weakness may be women's hidden strength.

This is considered to be a reactionary view, I learned to my sorrow when I proposed it to a female executive at RCA whose special interest is the careers of professional women. In response to my remark, she said, "I know what you mean well, but to tell a woman engineer that she has female intuition is like telling a black that he has rhythm."

Inevitably it occurred to me that anyone wondering why women do not become engineers would be well advised to learn something about the few women who *do* become engineers. So one day I took myself to the Engineering Societies Building, a large stone-and-glass structure overlooking the East River near the United Nations in New York City. In this stately edifice are housed most of the major professional societies that represent American engineers. On the third floor, past the imposing offices of the Engineering Foundation and the American Association of Engineering Societies, there is a single room that serves as the home of the Society of Women Engineers. The society, founded in 1959 by 50 women engineers, has grown from a membership of just a few hundred in 1970 to more than 9,000 in 1980. Still, compared to the other engineering societies, it seems pitifully small.

During my visit I browsed through a pile of career guidance pamphlets, newsletters full of recruiting ads from duPont, Boeing, Ford, and IBM, and also a booklet telling about the society's achievement award, given annually since 1952. The winners of this award are talented women who have made contributions in many fields: solar energy, circuit analysis, metallurgy, missile launchers, rubber reclamation, computers, fluid mechanics, structural design, heat transfer, radio-wave propagation, and so on. Their undeniable ability adds poignancy to the fact that they and their fellow women engineers are so few that their overall contributions to the profession have been, in essence, negligible.

In some of the society's literature I discovered a series of autobiographical essays prepared by society members. In each of these life-stories there was evidence of relatively humble family origins and of success earned through struggle. I also came across photographs of student-chapter members, smiling young women, mostly from the Midwest, who seemed—was it my imagination?—not at all like the sophisticated young women I had met at Smith.

Of course, the students at Smith do not study engineering. Neither do the students—male or female—at Harvard and Yale, which venerable institutions closed their professional schools of engineering years ago (although they still have some courses in engineering science), and neither of which deigned to respond to a recent statistical questionnaire from the Society of Women Engineers. All the circumstantial evidence I could garner served to reinforce my ideas about the class origins of the problem.

Wanting more information, I visited Carl Frey, executive director of the American Association of Engineering Societies, that organization of organizations to which most of the major professional engineering societies belong. In his position at the top of the organization pyramid, Frey has long lived with the many discontents and disputes endemic to the sprawling, variegated profession: four-year colleges versus five- and six-year programs (what constitutes a professional education?); state licensing (is an engineer a professional without it?); salaries (why do lawyers make so much more than engineers?); prestige (why do scientists get all the credit for engineering achievement?); leadership (why are there so few engineers in elective office?); conservatism of the self-employed versus radicalism of the hired hands; conscience, responsibility, the environmental crisis. Frey could not survive in his position without a genial disposition and a calm sense of history. From his point of view, women in engineering is just one more problem that the profession will cope with in due time.

"I wouldn't get hung up on any fancy theories about class," Frey said, after I outlined my hypothesis. "It's harder and harder to tell who comes from what class, and things are changing so fast that I wouldn't rely on any old statistics you might have seen about the social origins of engineers."

"Well, how do you explain it?" I asked. "Why aren't more bright young women getting into engineering?"

"I think that it has to do with their perception of power. These kids today—the bright girls particularly—they want to be where the action is, where the sources of power are. They don't see engineers as the ones who have the say in our society. And, let's face it, to a great extent they're right. We may have the knowhow, but we don't have the power."

Perception of power. The phrase kept going through my mind. It had a nice ring to it, and it had the ring of truth as well. It did not seem to contradict my ideas about class so much as to encompass them, for what is the origin of class structure if not the desire to perpetuate power?

Every engineer knows that the profession is relatively powerless. Engineers do not make the laws; they do not have the money; they do not set the fashions; they have no voice in the media. It is one of the most irritating ironies of our time that intellectuals constantly complain about being in the grip of a technocratic elite that does not exist.

To the extent that today's young women are not fooled by such nonsense, they are deserving of credit. But if intelligent, energetic women reject engineering because of an all-consuming desire to sit on the thrones of power, then woe to us all in the age of feminism.

When the National Organization for Women was formed in 1966, its Statement of Purpose spoke of bringing women "into full participation in the mainstream of American society *now,* exercising all the privileges and responsibilities thereof in truly equal partnership with men." Yet judging from the way that most advantaged women are selecting their careers, they seem to be a lot more interested in the privileges than in the responsibilities. In this they are following the lead of those males who appear to be in control of our society–the lawyers, writers, politicians, and business managers. This is all very well, but somebody in our society has to design, create, fabricate, build–to *do.* A world full of coordinators, critics, and manipulators would have nothing in it but words. It would be a barren desert, totally devoid of *things.*

Feminist ideology, understandably adopting the values of the extant, i.e., male, establishment, is founded on a misapprehension of what constitutes privilege. The feminist leaders have made the deplorable mistake of assuming that those who work hard without public recognition, and for modest rewards, are necessarily being exploited. "Man's happiness lies not in freedom but in his acceptance of a duty," said André Gide. When the duty turns out to be work that is creative and absorbing, as well as essential, then those who had been patronized for being the worker bees are seen to be more fortunate than the queen.

Studies have shown that young engineers, women as well as men, pursue their career because it promises "interesting work." This is more important to them than money, security, prestige, or any of the other trappings of power. They seem to recognize that a fulfilling career does not have to consist of a continuous ego trip.

Although power, in the popular imagination, is identified with wealth and domination, there is another kind of power that lies beneath the surface of our petty ambitions, and that is the engineer's in full measure. It is the force that Henry Adams had in mind when he wrote of the dynamo and the Virgin. The power of the Virgin raised the medieval cathedrals, although, as Adams noted, the Virgin had been dead for a millennium and had held no real power even when she lived. For better or for worse, technology lies at the heart of our contemporary culture, and the technologist is akin to a priest who knows the secrets of the temple. In this sense, and in this sense only, those who speak of a technocratic elite are touching on a profound truth. Until women share in the understanding and creation of our technology—which is to say, until large numbers of women become engineers—they will suffer from a cultural alienation that ordinary power cannot cure.

The feminist movement means different things to different people. Many of its goals, such as mutual respect and equality before the law, can be achieved even if there are no women engineers. But the ultimate feminist dream will never be realized as long as women would rather supervise the world than help build it.

## 12

# The Gentle Sophistries
# of the Club of Rome

The trouble with defending technology is that it entails arguing with a lot of nice people. ¶"Why are you always picking on the good guys?" a philosophy professor asked me one day at an academic conclave. ¶"Because the good guys are always making impractical suggestions," I replied. ¶There is a fine point to be made here. Of course we need dreamers and poets. Without wit, without vision, without sermons, we would be beasts or savages or robots, or something loathsome that we cannot even imagine. But without analysis and realistic planning, we would quickly succumb to chaos. Man does not live by bread alone, but neither can he survive on dreams and prayers. The Lord toiled for six days, and reviewed his handiwork on the seventh. This seems to me still to be a fairly good ratio by which to apportion our energies.

We cannot compartmentalize our lives, nor should we if we could. Our toil, ideally, is imbued with the inspiration of the sabbath; but the noble sentiments of the sabbath will not do the hard work that needs doing during the week. Only men and women, using technology–or rather, being technological–can do that work. Also, it requires taking the world pretty much as we find it, and moving on in modest advances. We cannot have political revolutions or spiritual upheavals at too frequent intervals if we are going to garner harvests and rebuild cities. Proclamations and manifestos cannot provide food and shelter, much less the excess wealth that makes art and philosophy possible in the first place.

This is why, as an engineer, I find myself exasperated so often by the "good guys," with their impractical suggestions and their extravagant preaching.

Wherever wishful thinking becomes the dominant mode of perception, the pragmatic technologist sees a source of potential danger to our society. It is for this reason that, since its inception, I have been interested in and apprehensive about the Club of Rome. For if ever good intentions and benevolent exhortation have been institutionalized, it is in that unique international organization. On the one occasion that they gathered together in the United States, I was anxious to see their operation at first hand.

When I checked into my Philadelphia hotel the night before the opening of the Club of Rome's 1976 meeting, a small green light on the wall was flashing insistently. A sign under the light said, "Call for Message." When I called, the operator said, "There is no message." I asked why the light was flashing. She replied, "The system is broken; we're trying to fix it." During the next few days, I thought often of that light.

Clearly, the global system in which we live is malfunctioning. Warning lights blink wildly all over the world. So it is reassuring to know that the members of the Club of Rome are dedicated to finding out what is wrong, and to prescribing a solution. They are a remarkable body of scholars, industrialists, and civil servants who give fresh luster to that worn-out phrase "men of goodwill." Their three-day meeting in the United States was a beautiful demonstration of moral concern. At the same time, judged by the standard of intellectual content, much of it bordered on the absurd. I had come to Philadelphia looking for a message about the future. Like the telephone operator at the hotel, the Club of Rome had no message, at least not one that I found intelligible. They filled the air with inspiring visions and noble proposals, hardly any of which came to grips with the problems of the world. At the close of the conference, I was left with a troublesome question: Can an intellectual disaster be a moral triumph?

The idea of holding the Club of Rome's meeting in Philadelphia is credited to Fulvio Oliveto, a member of the Philadelphia chapter of the Institute of Electrical and Electronics Engineers. Oliveto made his proposal to the club's chairman, the Italian industrialist Aurelio Peccei, who was enthusiastic. The idea was then taken to Philadelphia's prestigious Franklin Institute, and on to the First Pennsylvania Corporation, the city's leading financial institution, whose leaders had been looking for a suitable Bicentennial project. The corporation put up $240,000 to underwrite the cost of the conference, plus part of the cost of an exhibit on "futures" to be mounted at the institute. Surely the sponsors were not unmindful of the public relations benefits to be reaped from such an enterprise, but compared to all the foolish, costumed stagings of Revolutionary War battles and other embarrassing Bicentennial manifestations, the decision to invite the Club of Rome to Philadelphia stood as a model of intelligence and good taste.

The Club of Rome had never before held one of its full-scale meetings in the United States. It was coming to these shores bathed in a mystique almost without parallel for an organization so young and lacking in wealth, power, or constituency. Part of the club's fame is, undoubtedly, attributable to its elegant name. The word *club* has a social flavor that cannot be duplicated by *organization, institute,* or even *society. Rome* connotes imperial majesty, ecclesiastical grandeur, and continental sophistication. Add to this fortuitous choice of name an arresting first report seeming to predict the imminent end of the world, and you have the beginnings of instant renown.

The club was founded in 1968 at the home of Aurelio Peccei, following a leisurely luncheon at the Accademia dei Lincei in Rome. As Peccei tells the tale, the weather was balmy, the view from the academy was lovely, and the wine flowed freely. Thirty concerned citizens from ten nations had gathered to discuss humanity's ominous prospects, and to consider what a small group like themselves might do to improve them. The urbane aura of that occasion characterizes the deliberations of the club leaders to this day, making their meetings seem dilettantish, but, at the same time, establishing a mood of cordial civility that inspires faith in the possibility of rational solutions.

The club is "a queer animal," in Peccei's words, with no organization or staff, no formal minutes of its meetings, and practically no budget. "When we first tried to get support," says Peccei with a smile, "we got much support. Moral support." The club, however, has found a way around its lack of direct income. Its conferences, which take place almost every year, have been subsidized by the governments of Austria, Switzerland, and Canada, and by business groups in France, Japan, and now the United States. Its research projects are underwritten by governments, foundations, and corporations. The power of persuasion is clearly not the least of its members' talents. Chartered in Switzerland as a nonprofit association, the club's membership is limited to 100, and is drawn from all parts of the world, with the notable exceptions to date of Russia and China. Peccei is at pains to make clear that he has no grandiose plans for bureaucratic growth. The club's purpose, he says, is to act as a catalyst, to point out the nature of world problems, to propose alternative solutions, to alarm and enlighten governments and entire populations.

The web of global crises—technical, social, economic, and political—is labeled by the club the *problématique humaine.* The most significant characteristic of the *problématique* is an all-encompassing, interrelated complexity. The club maintains that such problems as food, population, resources, pollution, poverty, and so on can no longer be dealt with as identifiable, discrete matters, but must be considered as a dynamic maze of interacting phenomena. This does not appear to be an original thought—in fact, it seems downright obvious. However, the Club of Rome's great contribution was to try to be specific about what everyone knew to be generally true, to attempt to quantify and examine the forces at work. Seeking nothing less than a mathematical model for the whole world, it was inevitable that several club members should find their way in the summer of 1970 to MIT, where Professor Jay Forrester and his group were performing pioneering work in the field of systems dynamics. With financial support from the Volkswagen Foundation, an international team of researchers was put to work under the directorship of Dennis Meadows, and a year later the first report to the Club of Rome was ready. A popularized version of this report was published in March 1972 under the title *The Limits to Growth.*

The essence of the report was that exponential growth trends in population, industrialization, pollution, food production, and resource depletion threaten to bring us to the limits of global capacity within 100 years, resulting in catastrophe. No sooner had the ink dried (actually it never did dry, since the book has sold more than two million copies around the world and new editions are still being published) than the debate began between proponents of growth and no-growth. This issue, which made the Club of Rome world-famous overnight, has also proved to be something of an albatross. Peccei has tried vainly to explain that *Limits to Growth* was merely a first report *to* the club, and that it was not intended to be a statement of club policy. Club members have learned that it is easier to get your name into the newspapers than to get the story told to your satisfaction.

The argument about growth and no-growth seems to have generated much more heat than light. What Meadows said, after all, was that destructive growth is destructive, not exactly the sort of statement that should enrage reasonable people. Those who have attacked the report because it does not allow for the corrective actions people will take are coming very close to a tautology. People will indeed take action, not only because of automatic factors such as price changes (whose effect the report has possibly underestimated), but because of reasoned programs resulting from forecasts such as the report itself. In your warning, say these critics, you have neglected to consider that we might listen to your warning. Adding to the confusion is a lack of agreement about what exactly is meant by "growth." *The Limits to Growth* does not advocate a cessation of constructive activity, as some critics have assumed. Continuing technological advance, according to the text, will be "both necessary and welcome," as will "higher productivity," which could be "translated into a higher standard of living or more leisure or more pleasant surroundings for everyone." In short, the public debate, while not entirely without substance, proved to be an emotional argument between worried advocates of planning on the one hand and mildly optimistic advocates of laissez-faire on the other.

Members of the Club of Rome, although professing dismay at all the confusion and tumult, could not have helped but feel that *The Limits to Growth* was a success beyond their wildest dreams. The controversy it sparked had inspired the very debate it was the club's aim to encourage.

Then, suddenly, the entire picture changed. This animated, essentially academic, colloquy was interrupted by an outraged clamor of protest from an unexpected source–the developing nations of the Third World. "How can you have the effrontery," they asked, "to talk about limiting growth while we are starving and impoverished, just planning to embark on some growth of our own?" *The Limits to Growth,* they maintained, could only be viewed as part of a conspiracy to further subdue the exploited peoples, and the Club of Rome, as its name implied, was obviously an elitist agent of the imperialistic West.

The good and gentle members of the club were shocked and abashed. Certainly they had not intended to slight any of their brothers on this planet. They resolved to make amends. In so reacting they were already expressing the moral compassion and intellectual chaos that were to mark the Philadelphia meeting of April 1976.

The club did not abandon computerized forecasting on a global scale. A new world model was created by members Mihajlo Mesarovic and Edward Pestel, and stored in Mesarovic's computer at Case Western Reserve University. This model was more complex than the *Limits* model; it divided the world into ten distinct geographical regions and had data on different "levels" (individual, group, demo-economic, technology, and environment). It contained statistical information on approximately 100,000 relationships, such as birth rate to population growth, oil prices to fertilizer production, and capital stock to economic output. Scenarios could be played showing the probable impact of various alternative policies in the fields of agriculture, economy and finance, industrial investment, energy, and population control.

Work with this model formed the basis of the second report to the Club of Rome, published in 1974, entitled *Mankind at the Turning Point.* At the Philadelphia meeting, which was called "New Horizons for Mankind" (the titles begin to pall), one session was devoted to a report by Mesarovic and Pestel on use of the model as an alternative policy tool. At Case Western, Mesarovic was studying how alternative U. S. policies might affect the global food crisis. Other projects were underway in Germany, Iran, Venezuela, and Egypt. The progress reports on these projects were, I thought, the most substantial and interesting part of the conference. I heard some knowledgeable people complain that the model contained assumptions that were unwarranted, but to all such criticism Mesarovic and Pestel responded that they learned by doing, and were keeping the model "open" for modification.

If the club had restricted itself to improving such policy tools, advocating their use, and publicizing the results obtained with them, one could only report that they were performing a valuable service. Of course, this would mean diminishing headlines and a sense of frustration for Peccei and his colleagues, whose aim it is to prod the world continuously, vigorously, and in every conceivable fashion. It would also fail to satisfy the Third World critics of *The Limits to Growth*.

So the club embarked on two new ventures which were unveiled in Philadelphia: the RIO project (Reviewing the International Order), under the direction of the Nobel-Prize-winning Dutch economist Jan Tinbergen, and *Goals for Global Societies,* directed by the philosopher Ervin Laszlo. These two works may have stimulated enough controversy to satisfy the club's zest for perpetual agitation, but they also came close to damaging, permanently, the club's reputation among clear-thinking people.

The RIO report contains much solid information, and reflects devoted consideration of world economic problems, but it is, I believe, fatally flawed. In brief, it proposes that the rich nations make gifts to the poor nations, with the objective of reducing the 13-to-1 ratio that exists between average income in the richest 10 percent of nations and the poorest 10 percent. The word *gift* is not used, to be sure. Various euphemisms are adopted. There should be "transfers" of fertilizers and "transfers" of funds for development, "compliance" by transnational enterprises "with host countries' plans," "subventioning" of the cost of technological knowhow, and so forth. In addition, the developed countries should assist the developing countries by reducing tariffs, easing immigration restrictions, and levying taxes to support a central world treasury. "Continuation of the study," says a document distributed by the club, "may well indicate that the very concept of nation–state is outdated."

It takes no great insight into human affairs to conclude that citizens of the wealthier nations may not be willing to make the sacrifices called for in the RIO report. The report maintains that redistribution of wealth is required in order to avert worldwide disaster. But the opposite argument can be made more compellingly. Impoverished masses are much less likely to cause trouble for us than developing nations, which are just beginning to feel their oats. Angola-like controversies can arise, of course, but the superpowers have developed ways of handling such confrontations. The brutal truth is that the poorest nations do not pose a substantial threat to our well-being. Knowing, however, that the emerging nations will emerge eventually, whether we want them to or not, we are seeking their goodwill. We need their raw materials, we would like to have them as markets, and we want them to fall within our sphere of influence. Also, although we are terribly selfish, we want to do what is *right*.

What the average American considers to be right tends to be expressed in the form of what American leaders consider to be politically feasible. The outer limit of such policy is defined by present aid programs, augmented by the plans which from time to time are discussed at international conferences on trade and development, but nothing vaguely resembling the extravagant demands of the RIO package. RIO is not a rational proposal. It calls for more charity than people are willing to give. In order to become effective, it requires nothing less than a change in human nature. It is a sermon masquerading as a study.

If we protest, however, the Club of Rome is ready for us. It quickly brings Ervin Laszlo on stage with *Goals for Global Societies*. Human nature can change, says Laszlo. "Our researches show that the inner dimension of all major nations and cultures is capable of creative and humanistic transformation." His staff members came to this conclusion after studying polls and newspapers, conducting interviews, and in a variety of ways trying to capture the philosophical mood in different parts of the world. They claim to have evidence showing that there are humanistic goals that all people can accept and that will enable mankind to survive in a spirit of harmony.

Laszlo presented some specific findings: Americans believe that their level of consumption is immoral and that their politicians are not as forthright as they ought to be; in Western Europe the young are flocking to the ideals of the counterculture–utter honesty and self-limitation; in Eastern Europe there are socialist goals; in Japan there are indications that the aspirations of the average citizen are less materialistic than they were in 1973; in the Arab nations there is an urgency to catch up with the West, but a desire for something other than a consumer society; in Africa people are essentially religious; and so forth. "What we need," Laszlo said, "is an evolution of a new ethical consciousness."

The lights dimmed, and a slide was projected with the heading "The Required Transformation in Contemporary Values and Beliefs." At this point my notes become sketchy: "all religions . . . universal compassion . . . brotherhood . . . world solidarity." It was late in the day, and there was restless stirring in the hall as the third session of the conference drew to an end. Yet a sudden hush seemed to descend as Laszlo concluded. "We all have a moral obligation," he said, "to spur development of a sense of solidarity." If RIO was the sermon, then *Goals for Global Societies* was the closing hymn.

At the opening session Peccei had invoked the ethical and moral imperatives of the Declaration of Independence. Throughout the conference, speakers kept referring to the Bicentennial and the spirit of the American Revolution. Yet most conspicuously absent from the conference was the very pragmatism that characterized the American Founding Fathers. Jefferson and Franklin had ideals, but it never occurred to them to hang their hopes on anything as ephemeral as the "evolution of a new ethical consciousness." Their great achievement was to create a government for people who were imperfect, yet who wanted to live in freedom under a system of law. They were skeptical men of the Enlightenment.

The Club of Rome meeting was imbued with a very different spirit, a romantic neoidealism akin to that which prevailed in nineteenth-century Europe. Also, for all the lip service paid to the achievements of the United States, I sensed undercurrents of resentment and disapproval. The mood brought to mind not Philadelphia in 1776, but rather that same city in 1876, when the Centennial Exhibition attracted large numbers of European visitors. They crowded into the glass-and-iron Machine Hall to marvel at the many new mechanisms powered by the gigantic Corliss steam engine. American technology had come into its own, and astute observers could see that this portended significant changes for the human race. Europeans were impressed, but grudgingly so, patronizing the young nation as being technologically strong but woefully deficient in culture. A hundred years later this attitude persists. The 1976 Club of Rome meeting was, perhaps more than anything else, a genteel confrontation between the New World and the Old, between American pragmatism and European intellectualism.

The principal speakers at the meeting were all European. Two of them, Mesarovic and Laszlo, were American by citizenship, but born and educated in Europe. As for the Latins, Asians, Africans, Arabs and other non-Europeans, most of them, having been educated in the European tradition, shared the European mode of thought. The Americans were outnumbered about 10 to 1, and outtalked about 100 to 1. But when the final oratory died away, they seemed to have acquitted themselves very well.

For a while, things did not look promising for the image of

the host nation. The ugly American arrived in the person of Vice-President Nelson A. Rockefeller, who almost soured the affair beyond redemption. The opening banquet at the Franklin Institute was one of those festive occasions that impress even the jaded partygoer. The invited guests included not only the 80-odd conference participants from all over the world, about half of them Club of Rome members, but also a select group of Philadelphia citizens who had been invited by the sponsors. Everyone had been checked, and then checked again, by the Secret Service. The many agents with walkie-talkies and troopers with rifles heightened the dramatic tension. The director of the Institute walked serenely around the room, greeting as notable a collection of guests as his venerable building had seen in some time. All of a sudden the Vice-President was there, moving into the heart of the crowd, smiling, reaching out to shake hands. The feeling of power was electric.

Soon we were seated in the Benjamin Franklin Memorial Hall, under the huge white statue of Franklin by James Earl Fraser, dining on crown roast of lamb, and basking in the festive atmosphere. Then, after the coffee had been served, Nelson Rockefeller stood up and gave a harsh, crude, insulting speech that embarrassed everyone in the room almost beyond endurance. It was not a bad or uninteresting speech, as speeches go—a no-nonsense rebuke to unrealistic demands of Third World nations. However, before this group of benevolent humanitarians and invited guests, the effect was shocking.

It is to the credit of Nelson Rockefeller's reputation for responsibility that the almost universal assumption among the guests was that he had not seen the speech before he stood up to deliver it. Each of the three times he said that the most meaningful thing America could do to solve world problems would be to increase its own well-being, so as to serve as an example for others, he appeared to wince. When he called the Club of Rome naive for the second time, I felt that some speechwriter would soon be out of a job. He concluded by berating doomsday prophets and expressing total faith in the American people.

There was hardly any applause. Rockefeller shook hands with Peccei, who was flushed but grinning, trying to pretend that he had not been insulted. John Bunting, chairman of the First Pennsylvania Corporation, and host for the evening, gamely assured the Club of Rome members that they would have "equal time" the next day. The *Evening Bulletin* reported the event in the hockey parlance of the season: "Vice-President Nelson A. Rockefeller checked First Pennsylvania Corp. chairman John R. Bunting, Jr., into the boards last night. Bunting came up fast, saying it didn't really hurt."

None of the American members of the Club of Rome played an active role in the proceedings. Senator Claiborne Pell failed to make his scheduled appearance as co-chairman of the RIO session. Dennis Meadows, of *Limits* fame, showed up for the first day, but was gone on the second. Jay Forrester acknowledged applause from the dais at the final dinner, made one brief comment in public, and in private conversations grumbled about the "traveling circus" that the Club has become. However, those few invited American participants who did address the meeting spoke much sense in few words.

On the first day, during the RIO session, after Idriss Jazairy of Algeria had excoriated transnational enterprises, reciting a "litany of exploitation" and calling for their control by international "antitrust" legislation, G. William Miller, chairman of Textron (and later Secretary of the Treasury) was called upon to comment. "We must consider," said Miller softly, "the realities of human nature from the beginning of recorded time. The arguments we have heard will not convince the 'haves' to turn over their wealth to the 'have-nots.' This is not a negative comment. It is realistic. We must seek a confluence of self-interest." If there is a receptive climate, he continued, investment capital will flow from the rich nations to the poor. He suggested that we build with the

institutions that we have, trying to make the transnational corporations a force for good in the world.

I wondered if this straightforward approach might not bring
about a change in the tone of the meeting, but it was not to be.
The next speaker, Enrique Iglesias of Chile, responded defensively,
"Do we appear rhetorical and literary? Well, we are building a new
code of moral conduct."

The following morning Richard Gardner of Columbia Law
School tried to turn the meeting's attention to the limitations
imposed on all action by the imperfections of human beings. He
spoke of the ineffectual ways in which our political leaders function even when goals are not a matter of dispute. His wry humor
evoked little response.

On the final morning Arthur Stern, senior vice-president of
Magnavox, addressed the meeting briefly. Referring to the RIO
proposal to redistribute resources in the world, he said, "The populations of the wealthy countries will perceive such redistribution as
a sacrifice. We cannot postulate it as a categorical imperative. It
won't 'sell.' These are all wishes." It is not merely a question of
what is just, Stern tried to explain, as if talking to a child. We
must consider what is possible.

It is maddening to hear what seems to be pure common
sense, and to see that it is making no impression on the audience.
What was there about the people at this conference that makes
them immune to persuasion by evidence? "Everyone tries to dis-
courage us," Mesarovic told me. "We do not get discouraged."
Yes, but there is more to it than that.

One element was touched on the first day by the Indian
journalist Romesh Thapar, when he said that "those of us who
come to conferences are an elite who live luxuriously, copying
your ways." The members of this elite group in no way represent
the reality of life as it is lived by the masses in their countries. Nor,
on the other hand, do they represent the real power establishment

(a few maverick industrialists like Peccei notwithstanding). This point was made by another Indian, Professor Bacigha Singh Minhas of New Delhi, who said, "The ideology that we formulate may be tolerated by the upper class. But this is hypocrisy."

Members of the Club of Rome represent neither the proletariat nor the ruling classes, but that very thin layer of society which used to be called the intelligentsia. To a certain extent, this disqualifies them from speaking with authority about the future, for they represent nobody, and in any political upheaval they would be likely to disappear without a trace. One might even postulate that their interest in a world order stems in part from their frustration over the lack of a just order within their own homelands. During the three days of the conference there was no mention of the oppressive conditions that exist within the nations of so many of the participants. Not a word about political imprisonments, torture, corruption, violation of basic human rights, misappropriation of aid funds, and the rest. This may be in keeping with the etiquette of an international gathering, but it also reflects the willful blindness of those who are in love with their ideals.

During the final session of the conference a woman handed me a reprint of an article by Peccei entitled "The Humanistic Revolution." When I saw what it was, I felt like saying to her, "Dear lady, I have already read this, and I beg you to take all the copies of it that you can find and burn them quickly. Dr. Peccei is a good and kindly man, with wonderful talents for organizing and inspiring people. Tell him not to waste his time spinning these wild fantasies."

I picked up the article, which I had struggled through a few days previously. *"Something fundamental must be done to change human society and man himself. . . .* The challenge, in other words, is that of a quantum jump in human quality. Nothing less or different can suffice. And only a humane philosophy of life—*a new hu-*

*manism firmly established as the inspiration and guideline of society*–can generate and sustain this qualitative change." And on and on for eight magazine pages of closely set type.

All Club of Rome literature makes liberal use of italics to stress apocalyptic warnings and transcendental solutions. At this point I would like to italicize a sentence of my own. *We dare not trust the future of our children to any scheme that depends upon a change in human nature, particularly since the Club of Rome and others have shown convincingly that we cannot afford to wait for the millennium, but must plan and act promptly and continuously to meet the crises that confront us.* As Messrs. Miller, Gardner, and Stern told the conference, our only hope is to work with the people and institutions that exist. It is all very well to strive for the evolution of a new ethical consciousness. Who would not endorse such an effort? It is true, as Laszlo has said, that our attitudes are constantly changing, and sometimes such efforts have amazing success. Yet one thing has never changed, through the coming and going of great faiths, through the rise and fall of chieftains, emperors, doges, protectors, popes, and commissars, and that one thing is the struggle for wealth and power.

Such empirical reality does not impress the European intellectual. An Austrian graduate student tried to explain it to me once in a wine cellar in Salzburg: "You do not understand. We simply must have our theories."

It is all too easy to make fun of the implausible ideas of Peccei and his colleagues, and of the ornate sentences that filled the auditorium like music as the Club of Rome meeting drew toward its close. Ideas are wispy and have no reality until that sudden, unpredictable moment when they catch fire and explode. Then no one, least of all the person whose idea it was, can predict what will happen. From the witty conversation of Parisian salons, and some half-baked ideas of Jean-Jacques Rousseau, we can trace a line to the fall of the Bastille, Robespierre, and finally Napoleon. Ideas can be laughable, but they can also be frightening, particularly grandiose political ideas.

The ultimate expression of political intellectualism is found in the People's Republic of China. During the *Goals for Global Societies* session, Paul Lin, Professor of Asiatic Studies at McGill University, spoke of this phenomenon: "Freedom and welfare," he said, "are abstractions that mean nothing to the oppressed."

Beyond RIO and *Goals for Global Societies,* beyond all the Club of Rome visions of a new order, lies the reality of Communist China. It is the one place in the world where moral improvement is public policy. In our travail it beckons like easeful death, but not yet. Time enough for that if we fail.

At the end, the 1976 meeting of the Club of Rome seemed both ludicrous and frightening–and yet, as I said at the outset, inspiring. The same Arthur Stern who on the final day counseled the club members to return to the world of the possible was at the previous night's dinner comparing them to Diogenes and Jesus. "The deep faith in these men shines through," he said. "With all its shortcomings, the Club of Rome is unique."

At the press conference following the final session, Peccei and his executive committee members responded to questions that, while not hostile, were plainly skeptical. From the point of view of press coverage, the event was already a success, having received respectful front-page coverage in *The New York Times,* which in turn had brought representatives scurrying from *Time* and *Newsweek.* But Peccei, an evangelist to the end, was trying to persuade all the reporters present to carry the club's message forth continually to the public. He looked around the room, wistfully, wrinkling his brow like an aging Marcello Mastroianni. "Be a little naive," he said, "as we have been accused of being. It can be a better world."

A reporter asked, "Are you personally satisfied in your conscience that you are a model world citizen?" There was an embarrassed pause.

"We are not saints," answered Peccei. "But I will die with the belief that I did what I could."

# 13

## The Spurned Professional

ngineers, the creators of technology, are, by definition, a special target of the anti-technologists' assault. Thus, to gauge the full effect of the blaming of technology for our many ills, it is necessary to consider the impact of this trend upon the engineering profession. ¶The instinctive, first reaction of some individuals is anger. I am thinking of the mining engineer I met in Colorado who put a bumper sticker on his car that read: STOP MINING—LET THE BASTARDS FREEZE IN THE DARK. Then there is the nuclear engineer with his bumper sticker: MORE NUKES—LESS KOOKS. This response, while understandably human and an aid in venting frustration, exacerbates an already disagreeable situation. The angry engineer nourishes in himself a sense of petulant isolation, and helps to reinforce the antitechnologist's opinion that engineers are louts. ¶Many engineers are almost oblivious to the problem. They just want to be left alone. Some years ago, after I had delivered a speech at Brookhaven National Laboratory in which I discussed certain philosophical implications of technology, I was approached by a young engineer who spoke to me along the following lines: "Look, some of us were attracted to engineering because we're good at figures and because we're not good with people. We can do our work and solve most of the problems you give us, but please don't try to drag us into longwinded discussions about the meaning of life."

I was startled by this confrontation, which seemed to bear out the snide characterizations of engineers that one is likely to hear in the humanities departments of universities. Much as I advocate liberal education for engineers, I was struck by the young man's earnestness and concluded that there should be a place in our society for his specialized talent. If we accept the single-minded dedication of ballet dancers and other artists, we should be able to accept, however regretfully, the same characteristic in a number of our scientists and engineers. If technology were recognized by all segments of our society, including the intellectuals, as the splendid manifestation of humanity that it is, then such specialists would feel less isolated, and perhaps less reluctant to participate in other aspects of community life.

Some engineers respond to public hostility with a somewhat stoical paternalism. It is in the nature of things, I have heard some engineers say, that technically ignorant people will resent our knowledge and ability. It is just as useless to expect ordinary citizens to feel warmly toward us as it is to expect tenants to love landlords. We engineers had our time in the sun back in the good old days, before technology was seen to bring problems as well as benefits, and now we must simply put up with abuse from the uninformed, and do our work. Thus engineering becomes a matter of intellectual *noblesse oblige*.

To many engineers, the philosophical niceties of the anti-technology debate are less compelling than the economic practicalities. They are interested in making sure that engineering continues to get "its share of the pie." The image of the profession is seen to matter mainly to the extent that it affects the way budgets are prepared in government, academe, and industry. The concerns of this hardheaded group can range from basic questions of jobs and salaries up to the more lofty issues of industrial innovation and national productivity. When President Carter, in 1980, asked the Secretary of Education and the Director of the National Science Foundation to review American science and engineering education policies, the response of the engineering profession was mostly along the lines of "give us more money." The recommendations of the American Society for Engineering Education called for federal funding of fellowships for engineering faculty, funding for modernization of engineering school equipment—money, money, and more money.

Indeed engineering education in the United States has been deteriorating at the same time that it has been expanding. From 1969 to 1979, while the number of bachelor's degrees increased from 40,000 to 53,000 annually (32 percent), master's and professional degrees were growing only from 15,000 to 16,000 (7 percent), and doctoral degrees declined from 3,400 to fewer than 2,800 (−18 percent). Another distressing statistic: about 40 percent of the doctoral degrees have been awarded to non-United States citizens. This leads the ASEE to ask, "Where are the faculty coming from who will teach the classes of the 1990s?" The problem is stated, however, only in statistical and monetary terms. Nowhere in the report is the question asked, "What role does the spreading public hostility toward technology play in the decay of the engineering educational system?"

A review of these figures leads to other disturbing questions. What sort of a profession is this in which three-quarters of the members are merely graduates of four-year college courses? (And fewer than half take the trouble to obtain state licenses or to join professional societies?) Are all engineers authentic professionals, or only a few? Indeed, is engineering a profession or only a quasi-profession?

In 1978 I attended a conference at the National Academy of Engineering in Washington whose subject was "Education to Meet the Nation's Engineering and Related Technical Manpower Needs." It was an impressive occasion. Meetings in the Board Room of the Academy. Cocktails in the Great Hall of the Academy. Dinner in the Refectory. There were many fine speeches. Leaders of industry expressed a desire to find more young engineers with vision and sensitivity. Representatives of government spoke of the need for flexible and broadminded engineers in the public service. Leaders of the professional societies addressed the importance of noble concepts of professionalism.

But in the end, the academics took over the meeting with their charts and statistics, talking of supply and demand, budgets and tenure, as if the making of engineering graduates was some sort of assembly-line process. When a few of the participants protested, the president of one of the large Midwestern universities replied that his institution was merely turning out the product that was most in demand—competent technicians to fill specific job slots—and that all the grandiose speeches that had gone before verged on hypocrisy. His message was that the leaders of American industry and government do not want—indeed, will not tolerate—a large number of engineers who receive a rich, truly professional education. If there is any truth to that assertion, then it helps explain why our engineering schools are not graduating modern-day Leonardo da Vincis. I found that conference to be almost as illuminating as it was depressing.

The engineering profession today finds itself confronted with growing public suspicion, government neglect, and industry exploitation. It is still great fun to *do* engineering, but increasingly difficult to *be* an engineer. The reaction of many engineers ranges, as we have seen, from anger through apathy and stoicism to dismay verging on paranoia.

I wonder if it is not possible to formulate a more positive response. It has been said of Henrik Ibsen that he could not assert anything without suggesting the possible validity of its opposite; perhaps some such creative ambiguity is called for in the engineering community. What if neglect, lack of public support, even an element of antagonism were found to have a beneficial effect?

Human beings need resistance–something to work against. Children require it. Freed of restraint, they become anxious and disoriented. We can all testify, from our own experience and our observation of others, to the uses of adversity. Most of the great achievers–the great artists, the great statesmen, the great scientists–have overcome obstacles. In fact, they credit much of their achievement to the obstacles they found in their way. Resistance increases incentive. Challenge elicits new ideas and imaginative response. This is a cliché of biography.

What is true of individual lives is also true of entire cultures. Arnold Toynbee's great work has shown how civilizations have flourished in response to adversity. Not too much adversity, of course, as in the case of the Eskimos. But not too little, either. Which is Toynbee's point–and mine.

Something that seems obvious in individual lives, and in the evolution of cultures, appears not to be noted in connection with particular professions. Naturally, American technology could not have flourished without support from society. I believe it equally true that the engineering profession can benefit from the resistance and hostility of society.

In reflecting on this paradox, two images come to mind. The first is of an athlete as depicted in *The Inner Game of Tennis,* by W. Timothy Gallwey. "In tennis," says Gallwey, "who is it that provides a person with the obstacles he needs in order to experience his highest limits? His opponent, of course! Then is your opponent a friend or an enemy? He is a friend to the extent that he does his best to make things difficult for you. . . . So we arrive at the startling conclusion that true competition is identical with true cooperation."

The second image is of a Colonel Blimp-type character in a television commercial, staring blankly at the camera. All of a sudden a servant smacks him smartly on the cheek with a handful of aftershave lotion. "Thanks," says the startled colonel. "I needed that!"

In an unguarded moment, some engineers might agree. But then they would quickly point out that in fighting for a cause, one does not stop to grant points to the enemy. Opposition, it can be assumed, will always be provided in adequate amounts. Only by demanding everything and granting nothing does one end up with a reasonable share of the public pie.

But this attitude confronts us with another paradox. The banding together of engineers as a cohesive pressure group, if not done with discretion, is liable to increase public suspicion rather than allay it.

American engineers can be likened to an ethnic minority, and as such they are torn between two opposing objectives. One is to maintain their ethnic identity, and as an identifiable faction to get for themselves maximum rights, privileges, and recognition, and to make maximum contributions. The other objective is to assimilate, to maximize both reward and contribution by immersing themselves in what used to be called the melting pot and what we now call the mainstream. Some engineers, like some ethnics, are bitter, sensitive to every slight, and aggressive in seeking benefits for their own group. Others are all too anxious to deny their roots, and seek to escape from engineering into the world of business management.

Being part of an ethnic minority—as most Americans are, in one way or another—gives an individual a unique heritage and a special pride, while at the same time subjecting him to sometimes unwelcome social pressures. Ideally, this combination of factors serves to enrich life and to inspire creative performance.

For engineers, the disagreeable difficulties of today could conceivably be the animating spirit for the renaissance of tomorrow. This would, however, require a genial resilience that at the moment does not seem to be a dominant characteristic of the engineering profession.

# 14

# The Image Campaign

erhaps it was inevitable that some engineers, uneasy about the current status of their profession and uncertain about how best to improve the situation, should have sought the aid of another group of "experts"– the wizards of advertising. These clever manipulators lie in wait for the disconsolate, offering for sale the magic potion of cultivated reputations. ¶In a late 1979 issue of *Business Week* there appeared a peculiar advertisement. On the left-hand page was a large picture of an ambulance, lights ablaze, and above it the caption: "We civil engineers believe a carload of cantaloupes shouldn't come between an ambulance and its hospital." To the right was a small picture of a civil engineer standing in front of an insignificant-looking railroad overpass. Above this picture the caption continued:

"I could just imagine somebody with a coronary in an ambulance that has to wait for a seventy-two-car slow freight to pass. . . ." and paragraphs followed elaborating the social importance of engineering. The lengthy commentary ended: "For more information about how civil engineers serve people, write to the American Society of Civil Engineers."

The advertisement's more subtle message, I suppose, is this: Civil engineers, erstwhile heroes of a developing nation, are unhappy about having lost their appeal. No longer do they swagger through popular novels in high-laced boots or win the girl in movies about the building of the Union Pacific railroad. They are not to be found among the rock stars and politicians on television talk shows, nor in *People* magazine. What public attention is granted these days to science and technology is accorded for the most part to physicists and surgeons. And when the rare engineer does make news, he invariably represents one of the more glamorous branches of the profession: electrical, chemical, or aeronautical. The public takes for granted its railroads, highways, bridges, tunnels, airports, aqueducts, and sewers. It is bored by dams and skyscrapers and—as a result of environmentalism—increasingly hostile toward those who make them.

Civil engineers are understandably frustrated by the decline in status. Nonetheless, I was astonished to see this frustration transmuted into an advertising campaign. Traditionally, professional societies send slide shows to high schools and discreet public-service announcements to local radio stations. Occasionally they lobby, and even buy space in newspapers and magazines to speak out on issues. But they don't hire advertising agencies and pay large sums to the media in an attempt to buff up their public image. Certainly not the American Society of Civil Engineers: venerable (founded 1852), conservative, some would even say stuffy. When I learned that the society's board of direction had approved the campaign and announced it at their 1979 convention in Boston, my interest was piqued. How had this come to pass? And what are the broader implications of this curious enterprise?

At ASCE headquarters the staff was ready with answers to my first question. The campaign was undertaken simply because society members *wanted* it. According to James Shea, who was appointed to the newly created post of director of public communications, an extensive canvas of the members in 1973 set improved public relations as a high priority. Three years later, more than 20,000 of the society's 76,000 members were surveyed and, again, "gaining public recognition" was an activity most respondents thought deserved more attention. To this end, in 1977, the society's 125th year, a public relations campaign was approved, built around the theme "Civil Engineering–A People-Serving Profession."

But public relations proved inadequate, according to Mr. Shea. "Reporters are only interested in us when the roof falls down in the Hartford Civic Center. If you want to get your message across, you have to go directly to the public, and you can only do that through advertising."

However persuasive the argument, the society's staff and officers seemed nervous in anticipation of criticism. They had accumulated an array of reports and questionnaires that supported the advertising, and had prudently funded the campaign with an optional $4 contribution per member, collected along with the annual dues. About $100,000 had been collected, and that is the amount that was committed for the first phase. Not a large sum for an advertising campaign, Mr. Shea conceded, but enough to pay for two-page spreads in *Business Week, Engineering News-Record,* and a special executive-readership edition of *Time,* with something left over to pay for the society's float in the Tournament of Roses parade. (Started several years ago by the Los Angeles members, this anachronism has become an ASCE tradition.) It was hoped that more money could be raised so that *Newsweek* and *U. S. News & World Report* could be added to the list.

I asked if the recent U.S. Supreme Court decisions about professional advertising–ruling that it may no longer be prohibited by professional societies–had anything to do with the new campaign. Mr. Shea admitted that there was some connection. "A new category of advertising is coming into being," he said, "and we hope to set the tone."

As I left ASCE headquarters, I wondered how one goes about setting the right tone for braggadocio. Rationalize it as one will, to advertise is to boast, to puff—the antithesis of the diffidence traditional to engineering. What would the great civil engineers of the past think of this venture: John Smeaton, builder of the Eddystone Lighthouse, the first man to assume the title of civil engineer; John Rennie, who, after the opening of his Waterloo Bridge, said, "I had a hard business to escape knighthood"? What would be the comment of John Augustus Roebling, student of Hegel, creator of the Brooklyn Bridge? Surely these men would condemn the new campaign as a sordid business.

Yet modesty, admirable as it may be, is not the essence of honor. In fact, to set too much store by modesty, to seek an aloof dignity above all, verges upon the ignoble. Seen in this light, the advertising campaign is an act of courage for which the ASCE should be congratulated, having risked ridicule and charges of vanity in order to redeem the public value of a necessary discipline.

If professional self-acclaim is indeed to become a new category of advertising, the line between the praiseworthy and the fatuous will be very thin. Thus, a lot will depend on the performance of the "experts" to whom the professional societies turn for guidance.

The day after I visited ASCE headquarters, I called on the author of the campaign, Paul Lippman of Gaynor and Ducas, a small Manhattan advertising agency. He explained that the campaign is what is called in the trade "corporate advertising." Instead of trying to sell goods, its purpose is to affect the public's opinion of an institution. There is some financial motivation, to be sure, since civil engineers' salaries are to some degree related to the esteem in which civil engineering is held. But the main intent is to improve the image of the profession. This raises morale—a legitimate goal—and also serves to attract talented young people to the field.

"The first thing we look for," Mr. Lippman said, "is a creative rationale, a creative platform. I decided to stay away from big engineering projects. People can't relate to something like Hoover Dam. We have to be more specific and smaller in scale. So I thought we'd start with a single person, a real engineer, use his voice: 'We civil engineers believe. . . .' That has warmth, it's persuasive. Then I looked for a dramatic situation in which an *individual* is being helped by a civil engineer. For example, an ambulance that might be delayed by a train unless there's an overpass. Strong human interest."

Mr. Lippman admitted that he wrote the copy for the ad I saw in *Business Week,* but assured me that each of the engineers to be featured was a real person (nominated by his peers) who endorsed every word. "It could be you or your relative in that ambulance," Mr. Lippman said. "To be effective we've got to touch on the needs of individual people."

One might not care for Mr. Lippman's use of cantaloupes, but his point seemed reasonable enough. Even so, there was something about the ad, and those proposed to follow, that I did not like.

I walked a few blocks from Mr. Lippman's office to 57th Street and Madison Avenue, where excavation was proceeding for the new IBM building. Seventy feet below the surface a host of men and machines were carving an enormous hole into the rock, and I joined the people who were watching. An intricate pattern of steel and timber bracing supported the sidewalk on which we stood, suspended over the chasm. I turned west along 57th Street, past apartment houses, office buildings, stores, theaters, and art galleries. It was a beautiful, sunny day and the streets were filled with people strolling, shopping, sightseeing, or hurrying about their business. As I walked, I envisioned the maze of pipes and cables that lay beneath the pavement and the structural skeletons and mechanical networks in each of the buildings I passed. By the time I reached Carnegie Hall, its splendid tile-and-plaster ceiling suspended from 100-foot steel trusses, I knew what was wrong with the ASCE advertising campaign.

The essence of civil engineering lies not in what it can do for one person, but in what it does for the commonwealth. Ever since the first irrigation ditches were dug in Egypt, civil engineering has made it possible for large groups of people to live together. And ever since the building of the pyramids, civil engineering has provided the monuments and public works that inspire a sense of community.

If Mr. Lippman is right in his contention that people can no longer relate to Hoover Dam—or to Gustav Eiffel's tower, the Erie Canal, Yankee Stadium, the Golden Gate Bridge—then something lamentable has happened to people. Not so long ago, school-children studied the Seven Wonders of the Ancient World, and argued about what the seven modern wonders might be. Now it seems that no sense of wonder can be summoned at all.

An ambulance's unobstructed passage is the wrong image by which to define civil engineering. Yet it is difficult to fault the man who selected it. (Advertising, like technology, is one of those abstractions that people like to blame when the world disappoints them.) Mr. Lippman had assessed the public mood accurately: "What's in it for me?" But if civil engineers are to advertise, their proper role is to defy society's dispirited temper with symbols of civic grandeur.

If Mr. Lippman cannot find pictures of civil engineering projects to which a reader can "relate," let him start with *Twentieth Century Engineering,* the catalogue of a photography exhibit held at the Museum of Modern Art in 1964. And if, after looking at the breathtaking photographs of towers, vaults, bridges, and other works, he still is at a loss for a "creative platform," perhaps he can borrow a few sentences from the catalogue's introduction by Arthur Drexler:

> Engineering is among the most rewarding of the arts not only because it produces individual masterpieces but because it is an art grounded in social responsibility. Today we lack the political and economic apparatus that would facilitate a truly responsible use of our technology. But it may be that a more skillful and humane use of engineering depends on a more knowledgeable response to its poetry.

The present rebellion against materialism and size, however understandable its origins, shrivels the human spirit. People begin by searching for an inner peace and end by staring vacantly at sunsets, eating TV dinners in front of a flickering screen, and reassuring themselves that ambulances are standing by.

Civil engineering is an expression of group purpose and community pride—the counter-counterculture whose time is coming. It is sad to see the profession pandering to the very egoism it has a mission to transcend.

# 15
# Moral Blueprints

The most effective weapon in the arsenal of the antitechnologists is self-righteousness. Engineers can be stubborn and persuasive in debating technical issues, but they are pathetically vulnerable to assaults upon their honor. Like most professionals, they have cherished ethics and given it a central place in their philosophy. Thus, when confronted with the charge that they are lacking in ethical sensibility, they are thrown into confusion. ¶I discussed this problem in *The Existential Pleasures of Engineering;* but when I wrote that book in 1975, the campaign on behalf of technethics was in its early stages. Since that time there has been an explosion of concern. ¶Technethics? Yes, there is such a word. I discovered it when I had occasion to look something up in the 1973 *Britannica Book of the Year.* There it was, in a section called "New Words and Meanings," along with such terms as *air piracy, alternative society, counterproductive,* and *dingbat:*

> *Technethics:* The responsible use of science, technology, and ethics in a society shaped by technology.

The joining together of *technology* and *ethics,* two of the most supercharged words of the 1970s, has not been an etymological success (as was bioethics, for instance), but in all other respects the union seems to be thriving. Most people agree that this is all to the good. Ethics rejuvenated and applied to technology. Technology tempered by ethical considerations. It *sounds* good. It seems to be just what we need. Yet, as I look out upon a swelling tide of sermons, seminars, grants, articles, newsletters, conferences, professional codes, and academic courses, I wonder if anything is being accomplished. I am even troubled by the thought that what seems salutary might, upon inspection, prove to contain hidden elements of danger.

Of course, worries about the ethics of technology did not begin with the coining of the new word in 1973. Atomic weapons have been in our mind, and on our conscience, since 1945. By 1960, when Vance Packard called us wastemakers, we were already concerned about the world's diminishing natural resources. Soon Rachel Carson and Ralph Nader were expressing their anxieties, and the Club of Rome announced that environmental catastrophe was imminent. But it was not until 1973, with the oil embargo and the energy panic, that concern about technology reached pathological intensity.

As for ethics, the 1973 Watergate hearings suddenly brought that musty word out of theological texts onto the front pages of the nation's press. Also, misgivings about the morality of the Vietnam War seemed to intensify, rather than diminish, when American participation ended in early 1973.

That same year, the Agnew bribery scandal revealed that several prominent civil engineers had been involved in illegal payoffs, thus embarrassing a branch of the profession that had always taken pride in its tradition of integrity. Other incidents had raised questions about the morality of people engaged in technological activities. In 1968, at the B. F. Goodrich plant in Troy, Ohio, test results were falsified so that faulty aircraft brakes might be accepted for use on the U. S. Air Force A7D. In March 1972, three engineers were peremptorily dismissed by the management of BART (San Francisco Bay Area Rapid Transit District) for speaking to members of the Board of Directors about inadequacies in the train control system being furnished by Westinghouse.

In a different era these episodes might have been forgotten, but in the political climate of the 1970s they were resurrected to become the canon of a professional folklore, discussed again and again by concerned engineers and observers of the engineering scene.

The elements of an explosive new movement were visible everywhere, but the critical spark seems to have come in the field of the life sciences. In October 1971, an international symposium was held at the newly established Joseph and Rose Kennedy Institute for the Study of Reproduction and Bioethics at Georgetown University. At the conclusion of the meeting, 21 of the participants issued a statement calling for federal funds to support research and teaching in the area of science and values. Usually such statements disappear into dignified obscurity, but this one had an effect that should hearten participants in conferences everywhere. It was reported prominently by the press and referred to at Congressional hearings. The message was heard at the National Science Foundation, where within a month a task force was at work studying what role the NSF might play in this area. Similar concerns were voiced at the National Endowment for the Humanities, and on February 20, 1973, the NEH and the NSF jointly issued a document:

## IMPORTANT NOTICE
## TO
## PRESIDENTS OF UNIVERSITIES AND COLLEGES AND DIRECTORS OF NONPROFIT ORGANIZATIONS
Subject: Proposals Involving Ethical and Human Value Implications of Science and Technology

Recently, there has been mounting interest throughout our society in the ethical and human value implications of science and technology. . . . Although the National Endowment for the Humanities and the National Science Foundation have in the past supported activities related to this subject, both foundations, individually and in collaboration, are now prepared, on a selective and limited basis, to consider fresh approaches in support of scholarly activities in this field. Such approaches may include research and other forms of scholarly investigation, together with conferences, colloquia, seminars and similar activities.

Interest and anxiety had been growing. Now the needed ingredient—government money—was at hand. In a remarkable display of interagency cooperation, the two government foundations have sponsored a joint activity which, outside of academic circles, does not seem to have received attention commensurate with its importance. It is not a big program as government programs go; the NSF's Ethics and Values in Science and Technology, funded at just under $2 million annually, awards about 20 grants each year, mostly to engineers, scientists, and social scientists. The NEH's Program of Science, Technology and Human Values, which supports the work of philosophers, historians, and the like, had (prior to severe budget cuts in fiscal 1982) an annual budget of about $3.5 million, and awarded about 50 grants. Interdisciplinary activity is encouraged, and a number of joint NEH–NSF grants have been awarded to projects that bring scientists and engineers together with humanistic scholars. Although most of the funds go to academics, substantial grants are also made to museums, scientific and engineering societies, and other nonprofit institutions.

This government initiative has had an impact far greater than its comparatively modest statistics might suggest. First, it funded those essential elements of a new discipline: newsletters and surveys. At Harvard it supported a quarterly *Newsletter on Science, Technology & Human Values* (now published jointly by the John F. Kennedy School of Government at Harvard and the Program in Science, Technology and Society at MIT). At Lehigh University it funded the communiqués of a program called Humanities Perspectives on Technology. Aid has been provided to encourage editors of college engineering magazines to consider ethical issues. Cornell University and the American Association for the Advancement of Science were both commissioned to survey academic programs in STS (Science, Technology, and Society–a new field must have its own shorthand). Judicious support of historical research, textbooks, bibliographies, conferences, faculty appointments, and fellowships has had a ripple effect, lending respectability to interdisciplinary activities (traditionally scorned in departmentalized academe), and spurring universities, foundations, and industry to commit additional funds to the cause. The Carnegie Corporation of New York, for example, awarded a grant to the Hastings Center to conduct a study of the teaching of ethics at both the undergraduate and professional school level. The Andrew Mellon Foundation helped to support Rensselaer Polytechnic Institute's Center for the Study of the Human Dimensions of Science and Technology. An anonymous donor gave $100,000 to the Illinois Institute of Technology to establish a Center for the

Study of Ethics in the Professions. When MIT inaugurated its Science, Technology and Society Program in 1977, STS was already a flourishing field; initial development grants from the Sloan, Mellon, and Hewlett foundations totaled $2.5 million.

The objectives of the new technethics are difficult to define with precision, and the program's emphasis is constantly evolving. In the life sciences, professional morality was paramount from the beginning. In engineering, however, the stress at first was on the social history of technology and on studies that would serve to close the "two cultures" gap–technology in literature, engineering and public policy, humanities for technologists, and the like. Engineering ethics has gradually been given more attention, and this trend seems to be accelerating.

In the years following the NEH–NSF *Important Notice,* events lent new urgency to the technethics crusade. The DC-10 catastrophe over Paris in March 1974 was caused by a faulty rear cargo door which, according to falsely certified records, was said to have been repaired. During a fire at the TVA Brown's Ferry nuclear power plant in March 1975, one reactor ran out of control for almost seven hours; safety improvements recommended by government inspectors had been deferred because of cost. In February 1976, three General Electric engineers resigned in protest over what they considered to be inadequate safety provisions in the design of nuclear power plants.

At the same time the business community was shaken by evidence of bribery in foreign lands; it was revealed that Congressmen had been bought by Korean lobbyists, intelligence agencies had engaged in illegal activities, and doctors had cheated on their Medicare billings. Even the image of scientific research was tarnished, as white mice were painted black.

Public indignation increased with each new scandalous revelation. Congress, along with most of the state legislatures, rushed to enact new ethics laws. A *U. S. News & World Report* editorial in the summer of 1976 was entitled "The Boom in Ethics," and early in 1978 in a front-page story, *The New York Times* reported that courses on ethics "have moved into the mainstream of American universities and professional schools."

Uneasy about the lead taken by academe and the government, practicing engineers hurried to join the crusade. In late 1974 the Engineers Council for Professional Development adopted a new code of ethics which was quickly endorsed by most of the major societies. In May 1975 six of these societies gathered in Baltimore to cosponsor a Conference on Engineering Ethics, and since that time there has scarcely been a meeting of engineers anywhere that has not had its token speech on the subject. The technical journals are full of earnest homilies.

The engineering societies have long had their codes of ethics, but these codes traditionally stressed gentlemanly conduct rather than concern for the public welfare. An engineer was to be honest and impartial; he was to avoid conflicts of interest; he was not to criticize his fellow professional; and mainly, he was not to compete for commissions on the basis of price. Although the engineers who promulgated these principles believed that they were in the best interest of society, in the new age the old codes began to look like the rules of a self-serving guild.

It is ironic that just as the engineering societies agreed to modify their codes to stress broader ethical considerations, trying to move with dignity to higher ground, they were sent sprawling by a devastating blow. In April 1978 the U.S. Supreme Court ruled that the National Society of Professional Engineers had violated the antitrust laws by prohibiting competitive bidding among its members. Although the decision should have been expected (a 1975 ruling had abolished minimum legal fee schedules set by bar associations), and its practical effect will probably be negligible (most consulting engineers will still be selected on the basis of considerations other than price), many engineers reacted with dismay. The decision showed that although the professions like to think that they establish their own codes of conduct, in the end it is the rule of law that prevails.

Competitive bidding, however, does not stand out as an issue of primary ethical significance. Neither does advertising or licensing or maintaining competence through continuing education—all matters that rightly concern engineering leaders but do not go to the heart of professional ethics. The quintessential questions relate to the first canon of the newly revised Code of Ethics of Engineers:

> Engineers shall hold paramount the safety, health and welfare of the public in the performance of their professional duties.

The code previously had said that the engineer will have "due regard" for the public. A proposed change to "proper regard" was rejected as not being responsive to the demands of the times; only "paramount" would do. Almost everyone seems to agree that this revised statement is worthy of approval. Nobody, however, seems to be clear about exactly what it means, whether or not it can be taught, and if so, how, and by whom?

Questions about the meaning of the clause were at first drowned out by the clamor raised about the teaching of it. The universities moved quickly to stake out their claim. In industry there was some breastbeating, but scarcely any action. The professional societies, for all their conferences and promulgation of codes, accomplished little. (Since the vast majority of so-called professional engineers are employees of large corporations, neither licensed nor beholden to any association of their peers, the societies find it difficult to make much headway.) So by default and delegation, as well as by dint of their own efforts, it was in the universities that the ethical revolution was concentrated.

Once it was determined that professional ethics was to be

taught on campus, the debate began about which particular academics should do the teaching. In some professional schools (law, business, and medical, as well as engineering), the faculties rejected intrusion from the outside, claiming that ethics should be handled as an integral part of the existing curriculum. In an increasing number of schools, however, the doors have been opened to theologians, social scientists, and, especially, philosophers. The philosophers moved briskly into the field not only because they considered themselves qualified, but also—no laughing matter—because they needed the work. Derek C. Bok, the president of Harvard, wrote that the time was ripe for developing interdisciplinary programs in moral education "since professional schools are beginning to recognize the moral demands being made on their professions, while philosophy departments are finding it more and more difficult to place their Ph.D.s in traditional teaching posts." [29]

There are many skeptics, both within academe and without, who argue that moral character is formed in the home, the church, and the community, and cannot be modified in a college classroom. Even Bok, while asserting that ethics can and should be taught in the university, concedes that "formal education will rarely improve the character of a scoundrel."

Whatever my reservations about this burgeoning academic phenomenon, I cannot agree with the skeptics on this account. Character may very well be formed at an early age, but what does this signify when so many good people do so many terrible things? Most of the deeds that we find to be morally deplorable are not performed by villains, but rather by decent human beings—in desperation, momentary weakness, or because of an inability to discern what is morally right amid the discordant claims of circumstances.

The determination to *be* good may be established at an early age, but we grapple all our lives with the definition of what *is* good, or at least acceptable. I see nothing inherently wrong with doing some of that grappling in the classroom.

It would seem that an ethics program for engineers is a prudent step for society to take, akin to an inoculation of gamma globulin. However, since the new engineering ethics is founded upon a deceptive platitude, there is a potential for harm if too many people invoke it unthinkingly, in articles and speeches, and worst of all in classroom lectures. The revised code, which accurately reflects the current mood of guilt and introspection, holds that the professional's primary obligation is to the public–"the ultimate client," according to one teacher of engineering ethics. The engineer is no longer to be guided by his particular client's wishes, his employer's instructions, or by his own creative imagination, as constrained by laws, regulations, and technical parameters. He must answer first to what his conscience tells him is best for the common good. Technethics, born of public outrage, ends by seeking solutions in private virtue.

If this appeal to conscience were to be followed literally, chaos would ensue. Ties of loyalty and discipline would dissolve, and organizations would shatter. Blowing the whistle on one's superiors would become the norm, instead of a last and desperate resort. It is unthinkable that each engineer determine to his own satisfaction what criteria of safety should be observed in each problem he encounters. Any product can be made safer at greater cost, but freedom from risk is an illusion.

"Know thyself," said the ancient Greeks. "It is the beginning of wisdom." I consider myself as ethical as the next person, but I do not see that society would be well served by my making technical decisions in accordance with my personal moral whims. In the design of buildings, for example, surely it would not be right for me to apply my own ideas of adequate safety–stairs, exits, sprinklers, and the like. In fact, I am legally and professionally bound to abide by publicly approved codes. I can exceed code requirements, of course, but then I begin to violate the moral precept that requires me to be economical. Hospital directors, for example, have objected to certain building safety standards which they say unnecessarily drive up the costs of health care to the detriment of all citizens.

And am I to be guided by conscience in establishing standards for safety in the workplace? Construction workers are subject to many hazards, some of which are unavoidable. Fortunately, insurance companies insist upon certain precautions. Labor unions make demands on behalf of their members. Then come the government agencies–the Department of Health, the Department of Labor, the Fire Department, OSHA, and others. I am well satisfied that the regulations are better protection for the workers than any paternalistic good will I might evince. The same concept applies to controlling exhausts from my equipment, or noise, or blocking the sidewalks during construction, or a thousand other matters concerning public welfare. As a professional, I abide by established standards (and perhaps I may serve on a committee that writes codes or sets standards). As a human being I hope that I deal adequately with each day's portion of moral dilemma. But between legality on the one hand and individual predilection on the other, there is hardly any room for the abstraction called engineering ethics.

It is precisely in the narrowing gap between law and individual taste that the technethicists want engineers to apply their moral sensibility. They argue that not every eventuality can be covered by code, particularly when new products are being developed, and not every product can be adequately reviewed by representatives of public agencies.

Even in the unmapped areas, however, reliance upon the moral bias of engineers is not a rational policy. Consider the design of a new automobile: some engineers may personally favor maximum provisions for safety; others may believe in saving weight in order to conserve fuel; others may put a high premium on economy. An appeal to ethical standards will not help define the "right" design.

Ethicists tend to stress the need for safety, choosing to ignore the benefits of economy. In designing cars, however, as in planning hospitals or solving any engineering problem, economy must be recognized as being practically and morally desirable. It means resources conserved and savings for the consumer. In a competitive market it does not mean additional profit for the manufacturer. (If markets are not competitive, the problem is clearly political, not technological.) Critics of the automobile industry used to say that the manufacturers made large, luxurious cars in order to maximize profit. Now they claim that efforts to make small, light, economical cars spring from the same greedy impulse. It is obviously easier for the nontechnical observer to complain about immorality than to face up to the fact that difficult choices have to be made, and that these choices cannot be made easier or better by appealing to ethics.

When three young women were killed in an accident involving a Ford Pinto, criminal charges were brought against the designers of the car. During the course of that landmark trial, one of the principal design engineers testified that he had given a Pinto to his daughter. This convinced me–and perhaps helped convinced the jury, who found the defendants not guilty–that the engineer, in his heart of hearts, believed that the car was designed with adequate safeguards. This does not necessarily mean that the car is as safe as society wants it to be. Good intentions and high moral standards do not help an engineer establish the limits of acceptable risk.

Such limits, where not established by laws and regulations, are defined by the way product liability cases are decided in the courts. Principles of financial accountability have been evolving for a number of years. Concepts of criminal culpability are also beginning to take shape. Although the designers of the Pinto were found not guilty, the incident must have impressed engineers and corporate leaders everywhere.

In sum, public-safety policies are properly established, not by well-intentioned engineers, but by legislators, bureaucrats, judges, and juries, in response to facts presented by expert advisers. Many of our legal procedures seem disagreeable, particularly when lives are valued in dollars; but since an approximation of the public will does appear to prevail, I cannot think of a better way to proceed. It would be a poor policy indeed that relied upon the impulses of individual engineers.

Some proponents of the new ethics have decried the increasing reliance on rules, saying that we cannot—and should not—regulate everything. This complaint ignores the fact that we do already have regulations covering almost every public aspect of life, including wearing clothes in the street, keeping dogs on a leash, loitering in public restrooms, and so forth. What is true is that these regulations could not be enforced if it were not for the good will of the citizenry. But such good will stems from the knowledge that the regulations do exist, have been established democratically, can be changed the same way, and that there is a mechanism, however imperfect, for punishing violators while protecting the rights of individuals. The existence of rules does not signify a failure of conscience, but rather an attempt to make explicit what might be called a collective conscience. Where technology is involved, the use of written guidelines is especially appropriate.

There have been a few cases in which scientists far in the forefront of some exotic field—splitting the atom or research in recombinant DNA—have had special problems of conscience that do not lend themselves to resolution by law. But such rare instances, interesting as they may be, should not dominate our thinking about the behavior of the average technologist. Neither should we confuse engineers with doctors, of whom we require a special compassion. The analogy between professions has been made too hastily.

It will be argued that the pledge to serve the public is not intended to transform each engineer into an independent review authority, but simply to serve as an ideal. Yet even as an ideal the precept is insidious. Although it is true that the public welfare is one of the responsibilities of all engineers (also bus drivers, chefs, television executives, air traffic controllers, teachers, and almost everyone else), it is also true that the public welfare should not be assigned in some vague way to the profession as a whole. Just as some lawyers dedicate themselves to protecting consumers or prosecuting criminals, so is there a noble tradition of engineers in public service who do the research, write the codes, and make the inspections that keep technology in check.

Engineering has a place for both the *creator* and the *guardian;* the dynamic tension between the two, crucial to social vitality, has been obscured by the shapers of the new engineering ethics. The guardians are not necessarily more altruistic than the creators. Some of them, to be sure, are genuinely committed to working for the common good. But the existence of an adequate cadre of public service engineers depends mainly upon the determination of the public to hire professionals whose assigned task it is to protect the public interest. (It is much easier for people to complain about engineering ethics than it is for them to properly fund regulatory agencies and public interest organizations.) Individual engineers, of course, can move back and forth between one role and the other. Their professional commitment is to bring integrity and competence to whatever work they undertake.

In addition, I believe that engineers have an obligation to devote some effort to public service, either through their professional societies or in some other appropriate forum. "I hold each man a debtor to his profession," said Francis Bacon, "from the which as men of course do seek to receive countenance and profit, so ought they of duty to endeavor themselves, by way of amends, to be a help and ornament."

By extension, I hold each professional a debtor to society at large. The engineering profession should strive to educate the public on technical issues, and to provide expert advice wherever the great debates take place. This does not mean, however, that one should bring personal bias to bear upon every engineering project. The ethically sensitive engineer will not, in his daily work, seek opportunities to show moral superiority, but will welcome guidelines within which he can energetically pursue his calling.

The new ethics does not stop with considerations of public safety, but goes on to hold the engineer accountable for the quality of life in this technological age. Engineers used to say that since they did the things that society commissioned, or at least applauded, they could not be held personally responsible for any adverse consequences. Although this is essentially true, they have recently been persuaded to say otherwise. Not wanting to be taunted for being mere cogs in the social machine, and enjoying the feeling of importance that comes with being called shapers of culture, engineers have agreed to don messianic robes. The president of Polytechnic Institute of New York, in a speech representative of the trend, has proposed a new precept for the profession: "We shall in general design our systems so as to enhance and glorify man, not dehumanize him." [30]

No one could quarrel with such an objective. But just as the ethicists have failed to distinguish between creators and guardians, so have they confused the functions of *solving problems* and *establishing goals*. The problem-solver cannot factor his personal fancy into each equation. He must operate within constraints and expectations set by those who commission his work.

There are ways in which engineers can (and should) enter the public arena to participate in establishing goals, but this is very different from filtering their everyday work through a sieve of ethical sensitivity. As a class, engineers have neither the power nor the right to plan social change. If they did, we would be well on our way to George Orwell's *1984*–except that engineers are no more agreed upon how to organize the world than are politicians, novelists, dentists, or philosophers. Should we make small cars or large? Risk oil spills to keep energy costs down? Accept the hazards of pesticides in order to feed hungry people? Stop building a dam and thus protect an endangered fish? These are political questions; it is pitiful and a little frightening to see citizens abdicate their responsibilities by assigning them to the realm of engineering ethics.

If one thinks that corporate decisions should be made by department consensus–featuring self-critical group meetings as in a Chinese commune–that is an interesting social theory. If one wishes to give tenure to all employees and to oblige corporations not to discriminate against mavericks and internal critics, that also is a position that can be defended. (We do, after all, have labor unions, constantly changing labor laws, employment contracts, and so forth.) But these ideas, which go to the heart of how human society is organized, should not be advanced under the guise of "engineering ethics."

Should professionals work only on projects that they, as citizens, approve? The new ethics implies as much. It sounds good to say that enlightened professionals should lead. But paradoxically, it is essential that professionals should serve. I once heard the famous musicians Isaac Stern and Eugene Istomin discuss this paradox as it bedevils performing artists. Stern believed that he was obliged to use his art to further his political beliefs, promoting "good" causes and boycotting "bad" ones. Istomin argued that a musician has a responsibility to perform wherever people want to hear music. Relating this debate to engineering, I tend to side with Istomin. It is generally agreed that each person is entitled to medical care and legal representation. Is it not equally important that each legitimate business entity, government agency, and citizen's group should have access to expert engineering advice? If so, then it follows that engineers (within limits of conscience) will sometimes labor on behalf of causes in which they do not believe. Such a tolerant view also makes it easier for engineers to make a living.

I do not mean to be cynical. The highest morality, I believe, starts, not with ethical maxims, but rather with a recognition of life's complexities. It follows that if technologists are to be humanistically sensitized, I would have them study literature and history. The humanities are most true to themselves when they stress the pulsating diversity of life rather than the search for moral imperatives. As one professor of philosophy has put it, perhaps with unintended whimsy: "The state of the art in moral philosophy does not yet permit the divination of Ethical Truth, which must for now remain a matter of individual discernment." [31]

The tendency to express complicated problems in simplistic moral terms is often associated with liberalism. I hasten to say that I am equally apprehensive about the ways in which conservative forces are making use of the new ethical crusade. There are people in the business and professional communities who, under the guise of moral concern, see a welcome opportunity to repudiate the government controls they detest and fear. Witness, for example, American Viewpoint, Inc., an industry-supported, nonprofit, "educational corporation," located in Chapel Hill, North Carolina, whose directors include the president of the National Association of Manufacturers, the executive vice-president of the American Medical Association, the assistant to the president of the American Mining Congress, and other representatives of Establishment America. In 1976 this organization published a volume of essays entitled *The Ethical Basis of Economic Freedom,* copies of which were sent to all senators, 1,000 banks, all companies on the "Fortune 500" list, college presidents, and the deans of most graduate schools. Part of the funding for this distribution came from William E. Simon, then Secretary of the Treasury, later Chairman of American Viewpoint's board of directors, whose concluding essay in the book was entitled "A Challenge to Free Enterprise."

In November 1977 this organization took full-page advertisements in the nation's leading newspapers to announce the
founding of an Ethics Resource Center. The essence of the American Viewpoint message was expressed in one sentence from that
announcement: "When honesty and ethics sink down, centralized
authority and coercive regulations rise up." A special message was
addressed to the nation's professional organizations, spoken, one
can almost imagine, with a wink: "The more trust earned, the
fewer restrictions needed."

Thus from both the Left and the Right we find zealous ar-
mies marching under the banner of technethics. (Every army
marches under a holy banner, which is why I get edgy when I see
such banners on the horizon.) These forces are dangerous because
they attack from different directions the policy that is our best
hope: the painstaking development of rules and regulations equal
to the complexities of our technology.

The Clean Water Act, passed by Congress in 1972 was modified in 1977 by more than 100 amendments. Then the Environmental Protection Agency, under the watchful eye of industry,
environmental groups, and local governments embarked upon the
preparation of detailed regulations. The EPA policy is to encourage local planning and enforcement of waste treatment programs,
within federal standards that will eventually cover 21 categories of
industries and perhaps 400 subcategories.

The National Dam Safety Act of 1972 was passed after 125
lives were lost in the failure of a coal tailings dam at Buffalo Creek,
West Virginia. Inspection of 9,000 high-hazard dams, although
mandated by Congress in 1972, was not funded until 1977, when
39 students and teachers at a Bible college died in a dam failure at
Toccoa, Georgia. As the inspections proceed, under the aegis of
the U. S. Army Corps of Engineers, it has been found that more
than 30 percent of the dams require remedial action, some on an
emergency basis.

In mid-1980, the EPA issued disposal regulations for hazardous wastes, controlling the disposal of 40 million metric tons of wastes at 26,400 sites nationwide. The regulation initially covered 501 specific chemicals, 118 of which are considered extremely hazardous. We are now discovering that there are thousands of potentially dangerous substances in our midst. They simply must be tested, the often-confusing results debated, and decisions made by democratically designated authorities, decisions that will be challenged and revised again and again.

This is all an excruciatingly laborious business, but we cannot avoid it by appealing to the good instincts of engineers.

If the multitude of new regulations and clumsy bureaucracies has made life difficult for corporate executives, the solution cannot lie in promising to be good and eliminating the controls, but rather in consolidating the controls themselves and making them rational—also a never-ending task requiring intelligence and tenacity.

It is not surprising that the antitechnologists—intimidated by complexity and impatient with detail—should conclude that Utopia can be achieved through the moral reformation of engineers. (This is very much in keeping with the illusory concept of technocracy.) It is also understandable that some engineers—flustered by criticism and flattered by attention—should confusedly agree.

Political conflicts and philosophical debates, however, cannot be papered over by appeals to engineering ethics. To attempt to do so is irresponsible. Neither can the world's technological problems be formulated, much less solved, in terms of ethical rhetoric. In engineering, especially, good intentions are a poor substitute for good sense, talent, and hard work.

# 16

# Technology and
# the Tragic View

The blaming of technology, as we have seen, starts with the making of myths—most importantly, the myth of the technological imperative and the myth of the technocratic elite. In spite of the injunctions of common sense, and contrary to the evidence at hand, the myths flourish. ¶False premises are followed by confused deductions such as those that have been discussed in this book—a maligning of the scientific view; the assertion that small is beautiful; the mistake about job enrichment; an excessive zeal for government regulation; the hostility of feminists toward engineering; and the wishful thinking of the Club of Rome. ¶These in turn are followed by distracted rejoinders from the technological community, culminating in the bizarre exaltation of engineering ethics.

181

In all of this it is difficult to determine how much is simple misunderstanding and how much is willful evasion of the truth, a refusal to face up to the harsh realities that underlie life, not only in a technological age, but in every age since the beginning of civilization.

Out of the confusion has come a dialogue of sorts, shaped around views that are deemed "pro"-technology or "anti," "optimistic" or "pessimistic." I believe that we should be thinking in different terms altogether.

*House & Garden* magazine, in celebration of the American Bicentennial, devoted its July 1976 issue to the topic "American Know-How." The editors invited me to contribute an article, and enticed by the opportunity to address a new audience, plus the offer of a handsome fee, I accepted. We agreed that the title of my piece would be "Technology and the Human Adventure," and I thereupon embarked on a strange adventure of my own.

I thought that it would be appropriate to begin my Bicentennial-inspired essay with a discussion of technology in the time of the Founding Fathers, so I went to the library and immersed myself in the works of Benjamin Franklin, surely the most famous technologist of America's early days. Remembering stories from my childhood about Ben Franklin the clever tinkerer, I expected to find a pleasant recounting of inventions and successful experiments, a cheering tale of technological triumphs. I found such a tale, to be sure, but along with it I found a record of calamities *caused by* the technological advances of his day.

In several letters and essays, Franklin expressed concern about fire, an ever-threatening scourge in Colonial times. Efficient sawmills made it possible to build frame houses, more versatile and economical than log cabins—but less fire-resistant. Advances in transport made it possible for people to crowd these frame houses together in cities. Cleverly conceived fireplaces, stoves, lamps, and warming pans made life more comfortable, but contributed to the likelihood of catastrophic fires in which many lives were lost.

To deal with this problem, Franklin recommended architectural modifications to make houses more fireproof. He proposed the licensing and supervision of chimney sweeps and the establishment of volunteer fire companies, well supplied and trained in the science of firefighting. As is well known, he invented the lightning rod. In other words, he proposed technological ways of coping with the unpleasant consequences of technology. He applied Yankee ingenuity to solve problems arising out of Yankee ingenuity.

In Franklin's writings I found other examples of technological advances that brought with them unanticipated problems. Lead poisoning was a peril. Contaminated rum was discovered coming from distilleries where lead parts had been substituted for wood in the distilling apparatus. Drinking water collected from lead-coated roofs was also making people seriously ill.

The advancing techniques of medical science were often a mixed blessing, as they are today. Early methods of vaccination for smallpox, for example, entailed the danger of the vaccinated person dying from the artificially induced disease. (In a particularly poignant article, Franklin was at pains to point out that his four-year-old son's death from smallpox was attributable to the boy's *not* having been vaccinated and did not result, as rumor had it, from vaccination itself.)

After a while, I put aside the writings of Franklin and turned my attention to American know-how in the nineteenth century. I became engrossed in the story of the early days of steamboat transport.[32] This important step forward in American technology was far from being the unsullied triumph that it appears to be in our popular histories.

Manufacturers of the earliest high-pressure steam engines often used materials of inferior quality. They were slow to recognize the weakening of boiler shells caused by rivet holes, and the danger of using wrought-iron shells together with cast iron heads that had a different coefficient of expansion. Safety valve openings were often not properly proportioned, and gauges had a tendency to malfunction. Even well-designed equipment quickly became defective through the effects of corrosion and sediment. On top of it all, competition for prestige led to racing between boats, and during a race the usual practice was to tie down the safety valve so that excessive steam pressure would not be relieved.

From 1825 to 1830, 42 recorded explosions killed upward of 270 persons. When, in 1830, an explosion aboard the *Helen McGregor* near Memphis killed more than 50 passengers, public outrage forced the federal government to take action. Funds were granted to the Franklin Institute of Philadelphia to purchase apparatus needed to conduct experiments on steam boilers. This was a notable event, the first technological research grant made by the federal government.

The institute made a comprehensive report in 1838, but it was not until 14 years later that a workable bill was passed by Congress providing at least minimal safeguards for the citizenry. Today we may wonder why the process took so long, but at the time Congress was still uncertain about its right, under the interstate commerce provison of the Constitution, to control the activities of individual entrepreneurs.

When I turned from steamboats to railroads I found another long-forgotten story of catastrophe. Not only were there problems with the trains themselves, but the roadbeds, and particularly the bridges, made even the shortest train journey a hazardous adventure. In the late 1860s more than 25 American bridges were collapsing each year, with appalling loss of life. In 1873 the American Society of Civil Engineers set up a special commission to address the problem, and eventually the safety of our bridges came to be taken for granted.

The more I researched the history of American know-how, the more I perceived that practically every technological advance had unexpected and unwanted side effects. Along with each triumph of mechanical genius came an inevitable portion of death and destruction. Instead of becoming discouraged, however, our

forebears seemed to be resolute in confronting the adverse consequences of their own inventiveness. I was impressed by this pattern of progress/setback/renewed-creative-effort. It seemed to
have a special message for our day, and I made it the theme of my
essay for *House & Garden.*

No matter how many articles one has had published, and no
matter how much one likes the article most recently submitted,
waiting to hear from an editor is an anxious experience. In this
case, as it turned out, I had reason to be apprehensive. I soon heard
from one of the editors who, although she tried to be encouraging,
was obviously distressed. "We liked the part about tenacity and
ingenuity," she said, "but, oh dear, *all those disasters*—they are so
depressing."

I need not go into the details of what followed: the rewriting, the telephone conferences, the re-rewriting—the gradual elimination of accidents and casualty statistics, and a subtle change in
emphasis. I retreated, with some honor intact I like to believe,
until the article was deemed to be suitably upbeat.

I should have known that the Bicentennial issue of *House &
Garden* was not the forum in which to consider the dark complexities of technological change. My piece was to appear side by side
with such articles as "A House That Has Everything," "Live
Longer, Look Younger," and "Everything's Coming Up Roses"
(devoted to a review of Gloria Vanderbilt's latest designs).

In the United States today magazines like *House & Garden*
speak for those, and to those, who are optimistic about technology. Through technology we get better dishwashers, perma-
nent-press blouses, and rust-proof lawn furniture. "Better living
through chemistry," the old du Pont commercial used to say. Not
only is *House & Garden* optimistic, that is, hopeful, about technology; it is cheerfully optimistic. There is no room in its pages for
failure, or even for struggle, and in this view it speaks for many

Americans, perhaps a majority. This is the lesson I learned—or I should say, relearned—in the Bicentennial year.

Much has been written about the shallow optimism of the United States: about life viewed as a Horatio Alger success story or as a romantic movie with a happy ending. This optimism is less widespread than it used to be, particularly as it relates to technology. Talk of nuclear warfare and a poisoned environment tends to dampen one's enthusiasm. Yet optimistic materialism remains a powerful force in American life. The poll-takers tell us that people believe technology is, on balance, beneficial. And we all know a lot of people who, even at this troublesome moment in history, define happiness in terms of their ability to accumulate new gadgets. The business community, anxious to sell merchandise, spares no expense in promoting a gleeful consumerism.

Side by side with what I have come to think of as *House & Garden* optimism, there is a mood that we might call *New York Review of Books* pessimism. Our intellectual journals are full of gloomy tracts that depict a society debased by technology. Our health is being ruined, according to this view, our landscape despoiled, and our social institutions laid waste. We are forced to do demeaning work and consume unwanted products. We are being dehumanized. This is happening because a technological demon has escaped from human control or, in a slightly different version, because evil technocrats are leading us astray.

It is clear that in recent years the resoluteness exhibited by Benjamin Franklin, and other Americans of similarly robust character, has been largely displaced by a foolish optimism on the one hand and an abject pessimism on the other. These two opposing outlooks are actually manifestations of the same defect in the American character. One is the obverse, the "flip side," of the other. Both reflect a flaw that I can best describe as immaturity.

A young child is optimistic, naively assuming that his needs can always be satisfied and that his parents have it within their power to "make things right." A child frustrated becomes petulant. With the onset of puberty a morose sense of disillusionment is apt to take hold. Sulky pessimism is something we associate with the teenager.

It is not surprising that many inhabitants of the United States, a rich nation with seemingly boundless frontiers, should have evinced a childish optimism, and declared their faith in technology, endowing it with the reassuring power of a parent—also regarding it with the love of a child for a favorite toy. It then follows that technological setbacks would be greeted by some with the naive assumption that all would turn out for the best and by others with peevish declarations of despair. Intellectuals have been in the forefront of this childish display, but every segment of society has been caught up in it. Technologists themselves have not been immune. In the speeches of nineteenth-century engineers, we find bombastic promises that make us blush. Today the profession is torn between a blustering optimism and a confused guilt.

The past 50 years have seen many hopes dashed, but we can see in retrospect that they were unrealistic hopes. We simply cannot make use of coal without killing miners and polluting the air. Neither can we manufacture solar panels without worker fatalities and environmental degradation. (We assume that it will be less than with coal, but we are not sure.[33]) We cannot build highways or canals or airports without despoiling the landscape. Not only have we learned that environmental dangers are inherent in every technological advance, but we find that we are fated to be dissatisfied with much of what we produce because our tastes keep changing. The sparkling, humming, paved metropolises of science fiction—even if they could be realized—are not, after all, the home to which humankind aspires. It seems that many people find such an environment "alienating." There can never be a technologically-based Utopia because we discover belatedly that we cannot agree on what form that Utopia might take.

To express our disillusionment we have invented a new word: "tradeoff." It is an ugly word, totally without grace, but it signifies, I believe, the beginning of maturity for American society.

It is important to remember that our disappointments have not been limited to technology. (This is a fact that the anti-technologists usually choose to ignore.) Wonderful dreams attended the birth of the New Deal, and later the founding of the United Nations, yet we have since awakened to face unyielding economic and political difficulties. Socialism has been discredited, as was laissez-faire capitalism before it. We have been bitterly disappointed by the labor movement, the educational establishment, efforts at crime prevention, the ministrations of psychiatry, and most recently by the abortive experiments of the so-called counterculture. We have come face to face with *limits* that we had presumed to hope might not exist.

Those of us who have lived through the past 50 years have passed personally from youthful presumptuousness to mature skepticism at the very moment that American society has been going through the same transition. We have to be careful not to define the popular mood in terms of our personal sentiments, but I do not think I am doing that when I observe the multiple disenchantments of our time. We also have to be careful not to deprecate youthful enthusiasm, which is a force for good, along with immaturity, which is tolerable only in the young.

It can be argued that there was for a while good reason to hold out hope for Utopia, since modern science and technology appeared to be completely new factors in human existence. But now that they have been given a fair trial, we perceive their inherent limitations. The human condition is the human condition still.

To persist in saying that we are optimistic or pessimistic about technology is to acknowledge that we will not grow up.

I suggest that an appropriate response to our new wisdom is neither optimism nor pessimism, but rather the espousal of an attitude that has traditionally been associated with men and women of noble character—the tragic view of life.

As a student in high school, and later in college, I found it difficult to comprehend what my teachers told me about comedy and tragedy. Comedy, they said, expresses despair. When there is no hope, we make jokes. We depict people as puny, ridiculous creatures. We laugh to keep from crying.

Tragedy, on the other hand, is uplifting. It depicts heroes wrestling with fate. It is man's destiny to die, to be defeated by the forces of the universe. But in challenging his destiny, in being brave, determined, ambitious, resourceful, the tragic hero shows to what heights a human being can soar. This is an inspiration to the rest of us. After witnessing a tragedy we feel good, because the magnificence of the human spirit has been demonstrated. Tragic drama is an affirmation of the value of life.

Students pay lip service to this theory and give the expected answers in examinations. But sometimes the idea seems to fly in the face of reason. How can we say we feel better after Oedipus puts out his eyes, or Othello kills his beloved wife and commits suicide, than we do after laughing heartily over a bedroom farce?

Yet this concept, which is so hard to grasp in the classroom, where students are young and the environment is serene, rings true in the world where mature people wrestle with burdensome problems.

I do not intend to preach a message of stoicism. The tragic view is not to be confused with world-weary resignation. As Moses Hadas, a great classical scholar of a generation ago, wrote about the Greek tragedians: "Their gloom is no fatalistic pessimism but an adult confrontation of reality, and their emphasis is not on the grimness of life but on the capacity of great figures to adequate themselves to it." [34]

It is not an accident that tragic drama flourished in societies that were dynamic: Periclean Athens, Elizabethan England, and the France of Louis XIV. For tragedy speaks of ambition, effort, and unquenchable spirit. Technological creativity is one manifestation of this spirit, and it is only a dyspeptic antihumanist who can feel otherwise. Even the Greeks, who for a while placed technologists low on the social scale, recognized the glory of creative engineering. Prometheus is one of the quintessential tragic heroes. In viewing technology through a tragic prism we are at once exalted by its accomplishments and sobered by its limitations. We thus ally ourselves with the spirit of great ages past.

The fate of Prometheus, as well as that of most tragic heroes, is associated with the concept of *hubris,* "overweening pride." Yet pride, which in drama invariably leads to a fall, is not considered sinful by the great tragedians. It is an essential element of humanity's greatness. It is what inspires heroes to confront the universe, to challenge the status quo. Prometheus defied Zeus and brought technological knowledge to the human race. Prometheus was a revolutionary. So were Gutenberg, Watt, Edison, and Ford. Technology is revolutionary. Therefore, hostility toward technology is anti-revolutionary, which is to say, it is reactionary. This charge is currently being leveled against environmentalists and other enemies of technology. Since antitechnologists are traditionally "liberal" in their attitudes, the idea that they are reactionary confronts us with a paradox.

The tragic view does not shrink from paradox; it teaches us to live with ambiguity. It is at once revolutionary and cautionary. *Hubris,* as revealed in tragic drama, is an essential element of creativity; it is also a tragic flaw that contributes to the failure of human enterprise. Without effort, however, and daring, we are nothing. Walter Kerr has spoken of "tragedy's commitment to freedom, to the unflinching exploration of the possible." "At the heart of tragedy," he writes, "feeding it energy, stands godlike man passionately desiring a state of affairs more perfect than any that now exists." [35]

This description of the tragic hero well serves, in my opinion, as a definition of the questing technologist.

An aspect of the tragic view that particularly appeals to me is its reluctance to place blame. Those people who hold pessimistic views about technology are forever reproaching others, if not individual engineers, then the "technocratic establishment," the "megastate," "the pentagon of power," or some equally amorphous entity. Everywhere they look they see evil intent.

There is evil in the world, of course, but most of our disappointments with technology come when decent people are trying to act constructively. "The essentially tragic fact," says Hegel, "is not so much the war of good with evil as the war of good with good."

Pesticides serve to keep millions of poor people from starving. To use pesticides is good; to oppose them when they create havoc in the food chain is also good. To drill for oil, and to transport it across oceans is good, since petroleum provides life-saving chemicals and heat for homes. To prevent oil spills is also good. Nuclear energy is good, as is the attempt to eliminate radioactivity. To seek safety is a worthy goal; but in a world of limited resources, the pursuit of economy is also worthy. We are constantly accusing each other of villainy when we should be consulting together on how best to solve our common problems.

Although the tragic view shuns blame, it does not shirk responsibility. "The fault, dear Brutus, is not in our stars, but in ourselves. . . ." We are accountable for what we do or, more often, for what we neglect to do. The most shameful feature of the anti-technological creed is that it so often fails to consider the consequences of not taking action. The lives lost or wasted that might have been saved by exploiting our resources are the responsibility of those who counsel inaction. The tragic view is consistent with good citizenship. It advocates making the most of our opportunities; it challenges us to do the work that needs doing.

Life, it may be said, is not a play. Yet we are constantly talking about roles—role-playing, role models, and so forth. It is a primordial urge to want to play one's part. The outlook I advocate sees value in many different people playing many different parts. A vital society, like a meaningful drama, feeds on diversity. Each participant contributes to the body social: scientist, engineer, farmer, craftsman, laborer, politician, jurist, teacher, artist, merchant, entertainer. . . . The pro-growth industrialist and the environmentalist are both needed, and in a strange way they need each other.

Out of conflict comes resolution; out of variety comes health. This is the lesson of the natural world. It is the moral of ecological balance; it is also the moral of great drama. We cannot but admire Caesar, Brutus, and Antony all together. So should we applaud the guardians of our wilderness, even as we applaud the creators of dams and paper mills. I am a builder, but I feel for those who are afraid of building, and I admire those who want to endow all building with grace.

George Steiner, in *The Death of Tragedy* (1961), claimed that the tragic spirit was rendered impotent by Christianity's promise of salvation. But I do not think that most people today are thinking in terms of salvation. They are thinking of doing the best they can in a world that promises neither damnation nor transcendent victories, but instead confronts us with both perils and opportunities for achievement. In such a world the tragic spirit is very much alive. Neither optimism nor pessimism is a worthy alternative to this noble spirit.

We use words to communicate, but sometimes they are not as precise as we pretend, and then we confuse ourselves and each other. "Optimism," "pessimism," "tragic view"—these are mere sounds or scratches on paper. The way we feel is not adequately defined by such sounds or scratches. René Dubos used to write a column for *The American Scholar* that he called "The Despairing Optimist." I seem to recall that he once gave his reasons for not calling it "The Hopeful Pessimist," although I cannot remember what they were. What really counts, I suppose, is not what we say, or even what we feel, but what we want to do.

By saying that I espouse the tragic view of technology I mean to ally myself with those who, aware of the dangers and without foolish illusions about what can be accomplished, still want to move on, actively seeking to realize our constantly changing vision of a more satisfactory society. I mean to oppose those who would evade harsh truths by intoning platitudes. I particularly mean to challenge those who enjoy the benefits of technology but refuse to accept responsibility for its consequences.

Earlier in this chapter I mentioned the problems I encountered in preparing an article for *House & Garden,* and I would like to close by quoting the last few lines from that much-rewritten opus. The prose is somewhat florid, but please remember that it was written in celebration of the American Bicentennial:

> For all our apprehensions, we have no choice but to press ahead. We must do so, first, in the name of compassion. By turning our backs on technological change, we would be expressing our satisfaction with current world levels of hunger, disease, and privation. Further, we must press ahead in the name of the human adventure. Without experimentation and change our existence would be a dull business. We simply cannot stop while there are masses to feed and diseases to conquer, seas to explore and heavens to survey.

The editors of *House & Garden* thought that I was being optimistic. I knew that I was being tragic, but I did not argue the point.

# ACKNOWLEDGMENTS

Several members of the *Harper's* editorial staff, from 1976 to 1980, labored valiantly to improve my prose. I particularly want to thank Jeffrey Burke, Rhoda Koenig, Suzanne Mantell, Deborah McGill, and David Sanford. My debt to Lewis Lapham, the editor of *Harper's,* has already been noted in the Introduction.

Matthew Stevenson, an associate editor of *Harper's,* read the entire manuscript twice, and made invaluable suggestions. If I did not meet his exacting standards in every respect, I am sure that I improved the book in trying.

Were it not for the cajoling of Thomas Dunne, executive editor at St. Martin's, there probably would not have been a book at all. Tom is an irresistible luncheon partner; also a talented editor.

My wife, Judy, as twice before, endured with good humor a year of reclusive weekends and cancelled vacations. She also read the manuscript and made many sensible recommendations.

My sons proffered occasional help and frequent challenge. They have grown up to be my most stimulating critics— David because he thinks that I see imaginary Luddites behind every tree, and Jonathan because of what he fears technology is doing to the trees.

Virginia Crowley typed several drafts of the manuscript and helped in a hundred other ways, all the while doing her usual herculean work at Kreisler Borg Florman Construction Company.

Much of the statistical data, particularly in Chapter 4, was gathered by Neil S. Dumas, a senior scientist with the United States Government.

Finally, I want to record my debt to Hiram Haydn, who was the editor of *The American Scholar* from 1944 until his death in 1973. I never met Mr. Haydn, but when in 1960 he encouraged me to participate in an essay-debate in *The American Scholar* (part of which is included in Chapter 7) he emboldened me to address thereafter, not solely my fellow engineers, but a general audience as well. "This should be fun," he wrote to me at the time. It was, and it has been.

# NOTES

1. Robert Penn Warren, *U. S. News & World Report,* July 7, 1978.

2. Daniel J. Boorstin, "Tomorrow: The Republic of Technology," *Time,* January 17, 1977.

3. Anthony Lewis, *The New York Times,* May 26, 1980.

4. "The Talk of the Town," *The New Yorker,* April 16, 1979.

5. Ted Nelson, "Techno-Politics," *Penthouse,* October 1978.

6. Philip Handler, *Science,* June 6, 1980.

7. O. B. Hardison Jr., *Chronicle of Higher Education,* December 22, 1975.

8. Leo Marx, "American Literary Culture and the Fatalistic View of Technology," *Alternative Futures,* Spring 1980.

9. Michael Walzer, "The New Masters," *The New York Review of Books,* March 20, 1980.

10. Thorstein Veblen, *The Engineers and the Price System,* Harcourt

Brace & World, Inc. 1963, pp. 139-140. Originally appeared as a series of essays in *The Dial* in 1919.

11. *Engineering Education News,* May 1980.

12. Roy A. Medvedev, *On Socialist Democracy,* W. W. Norton & Company 1977, p. 300, published by Alfred A. Knopf, Inc., 1975.

13. Golightly & Co. International, Inc., New York, N.Y., Study of Chief Executive Officers, 1979.

14. *Korn/Ferry International's Executive Profile: A Survey of Corporate Leaders,* 1979.

15. *Wall Street Journal,* July 23, 1980.

16. Frederic J. Hooven, "The Decline of Detroit," *The Year 1979-1980,* Thayer School of Engineering, Dartmouth College.

17. John Dewey, *The Quest for Certainty,* 1929, Capricorn Books, G. P. Putnam's Sons, 1969, p. 262.

18. Susan Schiefelbein, "Is Nuclear Power a License to Kill?" *Saturday Review,* June 24, 1978.

19. Fred C. Shapiro, "Radiation Route," *The New Yorker,* November 13, 1978.

20. *Nuclear Power Issues and Choices,* Report of the Nuclear Energy Policy Study Group, Balinger Publishing Company, 1977.

21. Joseph Wood Krutch, *Human Nature and the Human Condition,* Random House, 1959, p. 99.

22. George Santayana, *The Life of Reason,* Charles Scribner's Sons, 1905-06, Volume I, p. 77.

23. Bertrand Russell, *My Philosophical Development,* Simon and Schuster, Inc., 1959, pp. 212-213.

24. Amory B. Lovins, "Energy Strategy: The Road Not Taken?" *Foreign Affairs,* October 1976.

25. E. F. Schumacher, *Good Work,* Harper & Row, 1979, pp. 25 and 27.

26. Martin Trow, "Some Implications of the Social Origins of Engineers," *Scientific Manpower* (National Science Foundation), 1958.

27. Ruth Schwartz Cowan, "From Virginia Dare to Virginia Slims: Women and Technology in American Life," *Technology and Culture,* January, 1979.

28. Marilyn Ferguson, *The Aquarian Conspiracy,* J. P. Tarcher, Inc. and St. Martin's Press, 1980, pp. 226 and 228.

29. Derek C. Bok, "Can Ethics Be Taught?" *Change,* October 1976.

30. George Bugliarello, "The New Engineering Ethics," speech delivered in Rochester, N.Y., June 5, 1978.

31. T. W. Lockhart, "Professional Societies and the Enforcement of Professional Codes," *Business & Professional Ethics,* Rensselaer Polytechnic Institute, Spring 1980.

32. John G. Burke, "Technology and Government," in *Technology and Social Change in America,* edited by Edwin T. Layton, Jr., Harper & Row, 1973.

33. Herbert Inhaber, *Risk of Energy Production,* 2nd ed. Ottawa, Ontario: Atomic Energy Control Board, AECB 119/REV-1, May 1978 (severely critiqued in June 1979 by the Energy and Resources Group at the University of California, Berkeley).

34. Moses Hadas, *A History of Greek Literature,* Columbia University Press, 1950, p. 75.

35. Walter Kerr, *Tragedy and Comedy,* Simon & Schuster, 1967, p. 107.

# INDEX

ABM debate, 37

Adams, Henry, 129

Advertising, 155–60; Court decisions about, 157

Airplane safety, 167

Akin, William E., 27

Alienation, 91, 92, 103–4

*All the Livelong Day: The Meaning and Demeaning of Routine Work*, 100

American Association for the Advancement of Science, 166

American Association of Engineering Societies, 127

American Engineering Standards Committee, 109–10

American National Standards Institute, 110, 116, 117

American Society for Engineering Education, 150

American Society for Testing and Materials, 108, 109, 111, 113

American Society of Civil Engineers, 108, 156, 157, 158, 184

American Society of Mechanical Engineers, 108

American Standards Association, 110

American Viewpoint, Inc., 178–9

Andrew Mellon Foundation, 166, 167

Antitechnology. *See* Technology, attitudes toward

Arafat, Yassar, 38

Army Corps of Engineers. *See* U.S. Army Corps of Engineers

Arthur D. Little, Inc., 19

*ASCE News*, 107

Asimov, Isaac, 2

*Atlas Shrugged*, 28
Atomic Energy Commission, 54, 63, 64
Atomic Industrial Forum, 61
Atomic power. *See* Nuclear power
Automobile: industry, 13–16, 21; safety, 173, 174
Autoworkers' strike, 99

Bacon, Francis, 175
BART system, 164
Battelle Memorial Institute, 18
Bazargan, Mehdi, 38
Bell, Daniel, 32, 33, 36
B. F. Goodrich plant, 164
*Blue-Collar Aristocrats*, 103
Boffey, Philip, 63
Bok, Derek C., 170
Boorstin, Daniel J., 7
*Brain Bank of America, The*, 63
Bridges, 184
Broomfield, John, 3
Building codes, 110. *See also* Regulation
Bunting, John, 143
Bureau of Consumer Protection, 118
Burnham, James, 30
*Business Week*, 155, 159
Butler, Nicholas Murray, 26

Caddell, Patrick, 80
Cairns, Walter, 18, 19

Carnegie Corporation of New York, 166
Carnegie Hall, 160
Carson, Rachel, 6, 163
Carter, Jimmy, 35, 62, 64, 150
CB radio, 21
*Changing Times*, 12
*Chicago Tribune*, 14
*China Syndrome, The*, 66
*Christian Science Monitor*, 15
Civil engineering, 45, 160. *See also* Engineering profession
Clean Water Act, 179
Club of Rome, 132–47, 163; founding of, 134; funding of, 134; Laszlo report to, 139–40; 1972 report to, 135–7; 1974 report to, 137; 1976 meeting of, 133, 140–7; purpose of, 134–7, 138; RIO report to, 138–9, 143–4
Coal-using power plants, 67, 187. *See also* Energy
Code of Ethics of Engineers, 169. *See also* Engineers
*Coming of Post-Industrial Society, The*, 32
Committee on Technocracy, 25–7
Commoner, Barry, 6
Conference on Engineering Ethics, 168
Congress. *See* U.S. Congress
Connecticut Yankee nuclear plant, 53–8; operation of, 56–8. *See also* Nuclear power
Consumer movement, 110

Consumer Product Safety Commission, 105
Corfam, 11, 18
Cornell University, 166
Corning Glass Works, 16
Council on Environmental Quality, 105
Cowan, Ruth, 124
Curtiss-Wright Corp., 15

"Dam Outrage," 43
*Dams and Other Disasters*, 43
David, Edward E., Jr., 37
DDT, 37
Defoliation, 37
Deltona Corporation, 49
Department of Energy, 64, 95
Department of the Interior, 95
Descartes, René, 76
Dewey, John, 50
Dickson, Paul, 99, 100
Diesel power, 85. *See also* Energy
Domhoff, G. William, 32
Donne, John, 92
Doolard, A. Den, 46
Douglas, William O., 43
Drexler, Arthur, 161
Duarte, José Napoléon, 38
Dubos, René, 192
Du Pont Company, 18

Easton, Ivan G., 112
Ecole Polytechnique, 38
Education policies, 150–1

Edwards, James B., 65
Eiffel Tower, 160
Eisenhower, Dwight D., 30
Electrical power: methods of distribution of, 85–7, 89. *See also* Energy
Electronic communication, 21
Electronic facsimile transmission (*fax*), 17
Ellul, Jacques, 6, 81
Energy: coal, 67, 187; debate, 85–9; diesel, 85; electrical, 85–7, 89; nuclear, *see* Nuclear power; oil, 15, 54; research, 87; solar, 85, 88, 187; wind-driven, 86
Energy Research and Development Administration, 64
"Energy Strategy: The Road Not Taken?," 85
Enfantin, Barthélemy-Prosper, 24
Engels, Friedrich, 24
Engineering profession: code of ethics of, 168–9; and design problems, 83, 173; duties of, 176; and ethics, 162, 164, 168–80; image of, 150, 155–60; and military planning, 45; numbers entering, 33, 150; political power in, 33–41; prejudice against, 122–3, 127; problems of, 127–8, 148, 150–4; and public works, 82–4; rewards of, 160–1; role of, $x$, 27; women in, 120–30

Engineers: attitudes of, 148–50, 152, 153–4; education of, 150–1; and ethics, 162, 164, 168–80; in government, 34–5; improving the image of, 155–9; in industry, 39–41; numbers of, 33, 121; political power of, 33–41; and public service, 175

Engineers Council for Professional Development, 168

Environmental issues; and Army Corps of Engineers, 47–51; flood control, 48; nuclear waste, 64; water supply, 47, 51, 82–3; waterway system, 47; wetlands, 48–9, 50

Environmental Protection Agency, 64, 105; policies of, 179–80

Erie Canal, 160

*Ethical Basis of Economic Freedom, The*, 178

Ethics. *See* Engineers, and ethics; Science; Technology

*Existential Pleasures of Engineering, The*, x, 71, 162

Federal Trade Commission, 106, 107, 112, 114, 115; rule on standards, 116–19

Feminist movement, 123–6, 129–30

Ferguson, Marilyn, 125

Fire, 182

First Pennsylvania Corporation, 133

Flaubert, Gustave, 76

"Flooding America in Order to Save It," 43

Fluidics, 16

Ford, Gerald R., 37, 62

Ford, Henry II, 15

Ford Motors, 15

Forrester, Jay, 135, 143

Franklin, Benjamin, 182, 183, 186

Franklin Institute, 133, 184

Frey, Carl, 127, 128

*Future of the Workplace, The*, 99

*Future Shock*, 7

Galbraith, John Kenneth, 32

Gallway, W. Timothy, 152

Gardner, Richard, 144, 146

Garson, Barbara, 100, 102

General Motors, 14, 15, 16

Georgetown University symposium, 164

Gide, André, 129

*Goals for Global Societies*, 138, 139–40

Goethe, J. W., 76

Gofmann, John, 60

Golden Gate Bridge, 160

*Good Work*, 97

Gould, Jay M., 31

Government. *See* U.S. Government

Ground-effect machines, 18

*Gulliver's Travels*, 95

Hackman, J. Richard, 100
Hadas, Moses, 189
Handler, Philip, 7
*Harper's* magazine, x
*Harvard Business Review*, 100
Harvard University, 166
Hastings Center, 166
Hegel, G. W., 191
Hersey, John, 7
Hewlett Foundation, 167
Holography, 18
Hoover, Herbert, 35, 110, 122
Hoover Dam, 160
*House & Garden*, 182, 185, 193
"Humanistic Revolution, The,"
    145
Humanities Perspectives on
    Technology program, 166
*Human Nature and the Human
    Condition*, 72

Ibsen, Henrik, 152
Ickes, Harold L., 43
*IEEE Spectrum*, 12
Iglesias, Enrique, 144
Illinois Institute of Technology,
    166
*Illusion of Technique, The*, 6
Industrial-safety codes, 110. *See
    also* Regulation
Industry, 109; engineers in,
    39–41; top positions in,
    39–40
Institute of Electrical and Elec-
    tronics Engineers, 12
Interagency Review Group: for

management of nuclear
    wastes, 64–5. *See also* Nuclear
    power
Inventions, 12, 18, 19
Istomin, Eugene, 177

James, William, 78
Jazairy, Idriss, 143
Job enrichment, 98–104; pro-
    grams, 98–101

Kafka, Franz, 76
Kahlil, Moustafa, 38
Kerr, Walter, 190
Killian, James R., Jr., 37
Koisch, Frank, 42
Korn/Ferry International study,
    39
Kranzberg, Melvin, 1, 2
Krutch, Joseph Wood, 71, 72,
    73, 75, 76

Lapham, Lewis, x, 67, 68
Laszlo, Ervin, 138, 139, 140,
    141, 146
Layton, Edwin T., Jr., 27
LDX machine, 17
Lehigh University, 166
LeMasters, E. E., 103
*Limits to Growth, The*, 135, 136,
    137, 138
Lin, Paul, 147
Lippman, Paul, 158, 159, 160
"Live from Studio 8H," 94

Lovins, Amory B., 85, 86, 87, 88, 89, 93, 95

McNamara, Robert, 36
*Main Street*, 90
*Managerial Revolution, The*, 30
*Mankind at the Turning Point*, 137
Marx, Leo, 9
Massachusetts Institute of Technology, 166, 167
Mazda Motors, 14, 15
Meadows, Dennis, 135, 136, 143
Medicine, 183
Medvedev, Roy A., 38
Mesarovic, Mihajlo, 137, 138, 141, 144
Military reactor sites, 63. *See also* Nuclear power
Miller, G. William, 143, 146
Mills, C. Wright, 31
Minhas, Bacigha Singh, 145
*Modern Temper, The*, 71
Mumford, Lewis, 6

Nader, Ralph, 54, 117, 163
National Academy of Engineering conference, 151
National Academy of Sciences, 37-8, 109
National Bureau of Standards, 109
National Dam Safety A t, 179
National Endowment fo the Humanities, 164-5

National Fire Protection Association, 110
National Organization for Women, 129
National Science Foundation, 37, 122, 164
National Society of Professional Engineers, 35, 168
"New Horizons for Mankind," 137
*New Industrial State, The*, 32
*Newsletter on Science, Technology & Human Values*, 166
*New Yorker, The*, 60
*New York Times, The*, 4, 8, 15, 168
Nietzsche, F. W., 68
Nixon, Richard M., 37
Nuclear power, 53-9; debate, 55, 66-9, 88; future of, 68-9; issue of safety of, 58-61, 66-9, 167; plants, 53-8, 66, 67-8; reactor sites, 63; and waste disposal, 63-6; waste products of, 61-3
*Nuclear Power Issues and Choices*, 67
Nuclear Regulatory Commission, 66
Nuclear weapons, 53, 62, 163; testing of, 37

Occupational Safety and Health Administration, 105
Office of Science and Technology Policy (OSTP), 37

Oil embargo, 15. *See also* Energy
Oliveto, Fulvio, 133
*Overskill*, 7

Packard, Vance, 163
*Pastoralism*, 9
Peccei, Aurelio, 133, 134, 135,
  138, 140, 143, 145, 146,
  147
Pell, Claiborne, 143
Pestel, Edward, 137, 138
Polaroid Corporation, 18
*Popular Mechanics*, 16
*Popular Science*, 14, 15
*Poverty of Power, The*, 7
*Power Elite, The*, 31
President's Science Advisory
  Committee, 37
Price, Don K., 31
Productivity, 101, 104. *See also*
  Work
Product standards. *See*
  Regulation; Standards
Proxmire, William, 43

Radiation, 58–61, 66. *See also*
  Nuclear power
Railroads, 184
Rand, Ayn, 28
Rasmussen, Norman C., 61
Rasmussen Reactor Safety Study,
  61, 66, 67
Rautenstrauch, Walter, 25, 26
Regulation: federal, of standards
  setters, 106–9, 112–19; need
  for, 105–6, 172–5, 179, 180

Rennie, John, 158
Rensselaer Polytechnic Institute,
  166
Reprocessing plants: for nuclear
  wastes, 62–3. *See also* Nuclear
  power
RIO project, 138–9, 143–4
Ritter, Donald, 35
*River Killers, The*, 43
Rockefeller, Nelson A., 142,
  143
Rodrigues, Benjamin-Olide, 24
Roebling, John Augusts, 158
"Role of Technology in Modern
  Society, The," 121
Romney, George, 36
Roszak, Theodore, 70, 81
Rotary engine, 11, 13–15, 20
Rousseau, Jean-Jacques, 93
Russell, Bertrand, 77, 78

SAAB, 99
Saint-Simon, Henri de, 23, 24
Santayana, George, 76
*Saturday Review*, 60
Schlesinger, James, 35–6
Schroeder, Robert J., 112, 113,
  114, 118
Schumacher, E. F., 81, 82, 84,
  85, 90, 93, 94, 95, 97, 98,
  102
Science: and concepts of man,
  72–4; education, 150; growth
  of, 30–1; and influence in
  government, 36–9; and val-
  ues, 75, 164–7, 174–5
*Scientific American*, 15

*Scientific Estate, The*, 31
Scientific world view: questioning the, 70–9
Scott, Howard, 26, 27
Sellers, William, 108
Shea, James, 157, 158
Simon, William E., 178
Sloan Foundation, 167
*Small Is Beautiful*, 81, 90, 95
Small-is-beautiful concept. *See* Technology: debate over "appropriate"
Smeaton, John, 45, 158
Society of Women Engineers, 126–7
Solar power, 85, 88, 187. *See also* Energy
*Spoon River Anthology*, 90
SST debate, 37
Standards, 172; and complaints about developers of, 114–17, 118–19; developers of, 110–12; history of, 108–10, 111; organizations, 109–10, 111, 116. *See also* Regulation
Steamboats, 183–4
Steiner, George, 192
Stern, Arthur, 144, 146, 147
Stern, Isaac, 177
Systems dynamics, 135

Taylor, Frederick W., 24
Technethics, 162–3. *See also* Technology: and ethics
*Technical Elite, The*, 31
Technocracy: concept of, 30–3, 36; definition of, 23; early

experiments with, 23–4, 30; movement, 24–7, 28, 30
"Technocrat," 28
*Technological Society, The*, 81
Technologists. *See* Engineers
Technology: and academe, 6, 7; attitudes toward, x–xi, 2–10, 28, 81, 92–3, 95, 148–52, 153, 156, 185–91; control of, 21–2; coping with the effects of, 182–5; debate over "appropriate", 82–96; and ethics, 163–76, 178–9; failures of, 12–18, 20–1; fear of, 4, 5, 7, 9; growth of, 30–1; and the humanities, 167, 178; influence of, on women, 124–5, 130; and the media, 4–5; myths about, 29, 33, 41, 181; need for, 131–2; new, 11–12, 19, 20, 87; problems of, 6, 88–9; and regulation, 105–6; social basis of, 53; theory of power elite in, 30–3; and the tragic view, 188–93
"Technology and Pessimism," 1, 8
"Technology and the Human Condition," 2
Television, 60
Terkel, Studs, 100, 101, 102
Thapar, Romesh, 144
3M Corporation, 18
Three Mile Island nuclear plant, 55, 67–8. *See also* Nuclear power
Tinbergen, Jan, 138

Titanium, 18
Toynbee, Arnold, 152
Toyo Kogyo, 14
Transistors, 17
Tribus, Myron, 20
*Twentieth Century Engineering,*
   160


UCLA Graduate School of Man-
   agement study, 40
Udall, Stewart, 43
Underwriters Laboratories, 110
United Auto Workers, 100
Urban homestead programs, 84
U.S. Army Corps of Engineers,
   42–51, 179; Civil Works Di-
   rectorate of, 44, 45; and
   environmental issues, 47–51,
   179; funding of, 44, 46; goals
   of, 49–51; projects of, 46–9
U.S. Congress, 46, 47, 49;
   Public Works Committees of,
   44–5
U.S. Government: agencies, 45,
   95, 105; engineers in, 34–5
U.S. Military Academy, 45
*U.S. News & World Report,* 167
U.S. Supreme Court, 168
Utilities. *See* Electrical power;
   Nuclear power: plants


Vaccination, smallpox, 183
Veblen, Thorstein, 25, 30
Vega plant strike, 99

Vietnam War, 163
Volvo, 99


*Wall Street Journal,* 15, 40
Wankel, Felix, 15
Warren, Robert Penn, 7
Washington, George, 35
Watergate hearings, 163
West Valley, N.Y., reprocessing
   plant, 62. *See also* Nuclear
   power
*Where the Wasteland Ends,* 81
Whitehead, Alfred North, 78
*Whole Earth Catalog, The,* 95
*Who Rules America?,* 32
*Who's Who in America,* 34
*Who's Who in Finance and Indus-
   try,* 39
*Who's Who in Government,* 34
Wind-driven power, 86. *See also*
   Energy
*Winesburg, Ohio,* 90
Winner, Langdon, 6
*Wisconsin Death Trip,* 90
Work: search for "fulfilling,"
   97–104
*Work in America,* 99
*Working,* 100


Xerography, 17
Xerox Corporation, 17
X-rays, 59, 60


Yankee Stadium, 160